Growing Graphene on Semiconductors

Growing Graphene on Semiconductors

edited by

Nunzio Motta | Francesca Iacopi | Camilla Coletti

PAN STANFORD PUBLISHING

Published by

Pan Stanford Publishing Pte. Ltd.
Penthouse Level, Suntec Tower 3
8 Temasek Boulevard
Singapore 038988

Email: editorial@panstanford.com
Web: www.panstanford.com

British Library Cataloguing-in-Publication Data
A catalogue record for this book is available from the British Library.

Growing Graphene on Semiconductors
Copyright © 2017 by Pan Stanford Publishing Pte. Ltd.
All rights reserved. This book, or parts thereof, may not be reproduced in any form or by any means, electronic or mechanical, including photocopying, recording or any information storage and retrieval system now known or to be invented, without written permission from the publisher.

For photocopying of material in this volume, please pay a copying fee through the Copyright Clearance Center, Inc., 222 Rosewood Drive, Danvers, MA 01923, USA. In this case permission to photocopy is not required from the publisher.

ISBN 978-981-4774-21-5 (Hardcover)
ISBN 978-1-315-18615-3 (eBook)

Printed in the USA

Contents

Preface   ix

1. The Significance and Challenges of Direct Growth of Graphene on Semiconductor Surfaces   **1**

N. Mishra, J. Boeckl, N. Motta, and F. Iacopi

1.1	Introduction	1
1.2	Direct Growth of Graphene on Si Substrates	3
	1.2.1 Laser Direct Growth	4
	1.2.2 Carbon Ion Implantation	5
	1.2.3 MBE Growth	7
1.3	Thermal Decomposition of Bulk SiC	7
1.4	Graphene on Silicon through Heteroepitaxial 3C-SiC	10
	1.4.1 Thermal Decomposition of 3C-SiC on Si	10
	1.4.2 Metal-Mediated Graphene Growth	12
1.5	Conclusions	15

2. Graphene Synthesized on Cubic-SiC(001) in Ultrahigh Vacuum: Atomic and Electronic Structure and Transport Properties   **27**

V. Yu. Aristov, O. V. Molodtsova, and A. N. Chaika

2.1	Introduction	27
2.2	Synthesis of Few-Layer Graphene	28
	2.2.1 Methods of Graphene Fabrication	28
	2.2.2 Growth of Cubic-SiC Epilayers on Standard Si Wafers	31
	2.2.3 Synthesis of the Epitaxial Graphene Layers on (111)- and (011)-Oriented Cubic-SiC Films Grown on Si Wafers	32
2.3	Synthesis and Characterization of Continuous Few-Layer Graphene on Cubic-SiC(001)	34
	2.3.1 Step-by-Step Characterization of SiC(001) Surface during Graphene Synthesis in Ultrahigh Vacuum	35

vi | *Contents*

2.3.2	Atomic and Electronic Structure of the Trilayer Graphene Synthesized on SiC(001)	41
2.3.3	Influence of the SiC(001)-c(2×2) Atomic Structure on the Graphene Nanodomain Network	53
2.4	Nanodomains with Self-Aligned Boundaries on Vicinal SiC(001)/Si(001) Wafers	55
2.4.1	LEEM and Raman Studies of Graphene/SiC(001)/4°-off Si(001)	55
2.4.2	Atomic and Electronic Structure of the Trilayer Graphene Synthesized on SiC(001)/2°-off Si(001)	56
2.4.3	Transport Gap Opening in Nanostructured Trilayer Graphene with Self-Aligned Domain Boundaries	59
2.5	Conclusions	63

3. Graphene Growth via Thermal Decomposition on Cubic SiC(111)/Si(111)   77

B. Gupta, N. Motta, and A. Ouerghi

3.1	Introduction	77
3.2	Epitaxial Growth of Graphene	78
3.2.1	Thermal Graphitization of the SiC Surface	78
3.2.2	Graphene Growth on Cubic SiC(111)/Si(111)	80
3.3	Surface Transformation: From 3C-SiC(111)/Si(111) to Graphene	82
3.3.1	Reconstructions of SiC(111)	82
3.3.1.1	Si-terminated face	82
3.3.1.2	C-terminated face	83
3.3.2	LEED and LEEM Characterization of the Transformation	85
3.3.3	STM Characterization: Atomic Resolution Imaging of the Transition	89
3.3.4	Atomic Structure Studies of Bi- and Multilayer Graphene	92

| | 3.3.5 | Improving the Epitaxial Graphene Quality by Using Polished Substrates | 96 |
| 3.4 | | Conclusion | 99 |

4. Diffusion and Kinetics in Epitaxial Graphene Growth on SiC

M. Tomellini, B. Gupta, A. Sgarlata, and N. Motta

4.1		Introduction	109
4.2		Evolution of Epitaxial Graphene Films as a Function of Annealing Temperature	111
	4.2.1	Evaluation of the Growth Rate in UHV	114
4.3		Growth Kinetics of Epitaxial Graphene Films on SiC	116
	4.3.1	Growth Kinetics under Ar Pressure	116
	4.3.2	Growth Kinetics in UHV	117
	4.3.3	Si and C Diffusion Process	118
	4.3.4	Kinetic Model of Graphene Layer-by-Layer Formation	121
	4.3.5	Kinetics of Graphene Islands with Constant Thickness	127
	4.3.6	Alternative Kinetic Models	132
		4.3.6.1 Terrace growth model	132
		4.3.6.2 Disk growth model	134
4.4		Conclusion	135

5. Atomic Intercalation at the SiC–Graphene Interface — 141

S. Forti, U. Starke, and C. Coletti

5.1		The Interface Layer	142
5.2		Hydrogen Intercalation	144
	5.2.1	How It Works	144
	5.2.2	Technical Details	144
	5.2.3	Quasi-Freestanding Monolayer Graphene	145
	5.2.4	Quasi-Freestanding Bilayer Graphene: Seeking a Bandgap	148
	5.2.5	The ABC of Quasi-Freestanding Trilayer Graphene	151
	5.2.6	Hydrogen Intercalation at the 3C-SiC(111)/Graphene Interface	156

viii | *Contents*

	5.2.7	Hydrogen Intercalation: Impact and Advances	157
5.3		Intercalation of Different Atomic Species	158
	5.3.1	The Intercalation of Ge Atoms at the Graphene/SiC Interface	160
	5.3.2	Electronic Spectrum of a Graphene Superlattice Induced by Intercalation of Cu Atoms	163
5.4		Conclusive Remarks	169

6. Epitaxial Graphene on SiC: 2D Sheets, Selective Growth, and Nanoribbons **181**

C. Berger, D. Deniz, J. Gigliotti, J. Palmer, J. Hankinson,
Y. Hu, J.-P. Turmaud, R. Puybaret, A. Ougazzaden,
A. Sidorov, Z. Jiang, and W. A. de Heer

6.1		Introduction	182
6.2		Near-Equilibrium Confinement-Controlled Sublimation Growth	183
	6.2.1	Multilayer C Face	185
	6.2.2	Monolayer C Face	188
	6.2.3	Monolayer Si Face	190
6.3		Selective Graphene Growth	192
	6.3.1	Masking Techniques	192
	6.3.2	Sidewall Facets	194
6.4		Large-Scale Integration	198
6.5		Conclusion	199

Index 205

Preface

Graphene, the wonder material of the 21st century, has not yet achieved the expected outcomes in terms of applications in nanoelectronics. This is not surprising, as large-scale graphene growth is still mostly limited to CVD on metallic foils, followed by graphene transfer to the semiconductor substrate required for electronic devices, which is cumbersome and difficult to automatize. Moreover, graphene is gapless, and this is still seriously limiting its applications.

This book is an attempt to dispel this pessimistic outlook, summarizing the latest achievements in the direct growth of graphene on semiconductors.

SiC is the ideal semiconductor for graphene growth, which is typically achieved by thermal graphitization. Through high-temperature annealing in a controlled environment, it is possible to decompose the topmost SiC layers, obtaining quasi-ideal graphene by Si sublimation. Graphene on SiC with record electronic mobilities has been demonstrated, opening the way for applications in nanoelectronics, by exploiting selective growth on patterned structures and gap opening by quantum confinement.

The book opens with a chapter on the significance and challenges of graphene growth on semiconductors, by Mishra et al., drawing a picture of the perspectives of this technology.

The three following chapters, by Aristov et al., Gupta et al., and Tomellini et al., respectively, are dedicated to an up-to-date analysis of the synthesis of graphene on SiC in ultrahigh vacuum. The fifth chapter, by Forti, Starke, and Coletti, is a review of the effect of atomic intercalation at the SiC–graphene interface, with an in-depth discussion of the doping effects and of the electronic properties of lateral superlattices. The sixth chapter, written by the de Heer and Berger group, reporting graphene growth on SiC by sublimation, summarizes the whole history of graphene growth on SiC by confined controlled sublimation, up to the latest developments in the growth of templated graphene nanostructures.

Preface

We hope that this book can be of inspiration to the many scientists striving to improve the growth of graphene on semiconductors and to the young researchers from industry and academia approaching this fascinating world. The developments sketched here show a promising outlook for the future, with many exciting outcomes on the horizon, from the artificial opening of a gap to the creation of 2D field-effect transistors with nanodimensions.

Nunzio Motta
Queensland University of Technology, Australia
Francesca Iacopi
University of Technology Sydney, Australia
Camilla Coletti
Istituto Italiano di Tecnologia, Italy

Chapter 1

The Significance and Challenges of Direct Growth of Graphene on Semiconductor Surfaces

N. Mishra,[a,d] J. Boeckl,[b] N. Motta,[c] and F. Iacopi[d]

[a]*Environmental Futures Research Institute, Griffith University, Nathan, 4111 QLD, Australia*
[b]*Materials and Manufacturing Directorate, Air Force Research Laboratories, Wright-Patterson AFB, OH 45433, USA*
[c]*Institute for Future Environments, Queensland University of Technology, 2 George Street, Brisbane, 4001 QLD, Australia*
[d]*School of Computing and Communications, Faculty of Engineering, University of Technology Sydney, Broadway, 2007 NSW, Australia*
francesca.iacopi@uts.edu.au

1.1 Introduction

In the past decade, fundamental graphene research has indicated several excellent electronic properties for graphene, such as ultrahigh carrier mobility ($\sim$200,000 cm^2/V·s), micrometer-scale mean free path, electron–hole symmetry, and quantum Hall effect [1–6]. Such

---

Growing Graphene on Semiconductors
Edited by Nunzio Motta, Francesca Iacopi, and Camilla Coletti
Copyright © 2017 Pan Stanford Publishing Pte. Ltd.
ISBN 978-981-4774-21-5 (Hardcover), 978-1-315-18615-3 (eBook)
www.panstanford.com

extraordinary properties, unmatched by any other conventional thin-film material, make it an extremely promising material for next-generation nanointegrated devices. Despite this, several fundamental challenges still lie ahead, before the introduction of graphene in nanodevices can be envisaged. One major challenge is the ability to confirm the outstanding reported properties for graphene grown over large areas onto appropriate substrates.

Since graphene was isolated the first time, in 2004 [7], several techniques have been demonstrated to produce high-quality graphene. The most common techniques are micromechanical exfoliation of single-crystal graphite, chemical vapor deposition (CVD) growth on transition metals and dielectric insulators, chemical reduction of graphite oxide (GO), carbon nanotube (CNT) unzipping, and high-temperature thermal decomposition of silicon carbide (SiC) [7–15]. Among these methods, the highest-performance graphene devices have been fabricated using mechanically exfoliated flakes. Carrier mobility in excess of $\sim$200,000 cm^2/V·s has been reported for suspended single-layer exfoliated graphene at room temperature [6, 16]. CVD growth is widely used to produce large-area (up to 30 inches) high-quality graphene on transition metal substrates [12].

However, the graphene layers produced in the ways described above need invariably to be transferred onto a semiconducting or insulating substrate for device fabrication. Unfortunately, for several compelling reasons, this transfer approach is not compatible with the commercial fabrication of actual nanodevices. First, the transfer of flakes tends to affect the quality of the graphene layer in terms of contamination and formation of detrimental folds and ripples, which can ultimately degrade the performance of the electronic devices [17, 18]. Moreover, it is unthinkable that the large-scale fabrication of micro- and nanodevices at the wafer level could be realistically achieved by transferring and positioning individual, ex-situ-grown, micrometer-size graphene flakes. In response to this bottleneck, researchers at IBM had recently demonstrated a methodology to achieve a full wafer-level transfer of a single graphene sheet [19]. However, even this approach will face challenges in terms of rippling, as well as low adhesion to the underlying substrate [20, 21].

To address these issues, numerous groups have focused on growing graphene directly on various semiconductor or insulating substrates, including silicon (Si), quartz (SiO_2), germanium (Ge), sapphire (Al_2O_3), magnesium oxide (MgO), aluminum nitride (AIN), boron nitride (BN), and silicon carbide (SiC) [10, 14, 22–30]. Direct integration of graphene on aforementioned substrates has great potential for numerous applications, such as interconnects, solar cells, superconducting material, high-frequency field-effect transistors (FETs), and optical modulators, as well as part of integrated heat sink structures for efficient removal of thermal dissipation [19, 31–36].

Among all substrates reported so far, graphene growth on SiC has gained by far the most interest in the scientific community due to the high quality of the obtained graphene. As a consequence, a great degree of understanding has been developed on the graphene on SiC growth and on the control of the properties of the grown graphene, as the following chapters will demonstrate. Despite being much more challenging, silicon as a substrate for direct graphene growth has also attracted great attention, given its technological importance in the semiconductor industry and the access it gives to the well-established silicon-based fabrication technology.

This chapter presents an overview of direct graphene growth on Si and SiC substrates. We first discuss the relevance and challenges of graphene growth on Si substrates. Thereafter, we provide a comprehensive scientific progress of graphene growth on silicon carbide (SiC) to date and evaluate its future perspective. Finally, we discuss the graphene growth from heteroepitaxial cubic silicon carbide (3C-SiC) on large-area Si substrates as the most promising method to achieve direct growth of graphene on silicon substrates. The chapter concludes with a brief discussion on the impact of graphene growth on Si and SiC in connection to future technology.

1.2 Direct Growth of Graphene on Si Substrates

In the last decade, several researchers have reported fabrication of Si-based graphene devices [31, 32, 37–40]. Most of them require postsynthesis transfer of the graphene onto a Si wafer, which is not suitable to present-day Si technology. Furthermore, the low adhesion energy of transferred graphene can significantly affect the device reliability [20]. To address these issues, various methods, such as laser direct growth [22], carbon ion implantation [23, 41, 42], and molecular beam epitaxy (MBE) [43, 44], have been reported to grow graphene directly on a Si substrate. In this section, we will briefly discuss these above-mentioned techniques.

1.2.1 Laser Direct Growth

In the case of laser direct growth method, first, a poly(methyl methacrylate) (PMMA) film is coated on a Si(111) wafer by spin coating [22]. The coated Si wafer is then covered by a quartz wafer, fixed on a sample stage, and placed in a vacuum chamber. After that, the sample is irradiated with a continuous laser beam (532 nm) of 3.1 W power. The laser beam provides thermal energy to the sample, which decomposes the PMMA film and melts the Si surface. The molten Si absorbs the carbon atoms released from the PMMA decomposition. Finally, the dissolved carbon atoms extracted from the Si melt upon cooling to form a few-layer graphene on the Si surface. Figure 1.1 shows optical images and Raman spectra of the laser-irradiated area on the sample with different laser-illuminated times, ranging from 1 s to 15 min [22]. It is clear from the image (Fig. 1.1a) that the PMMA is removed from the center in 1 s, although no visible change is observed in the bare silicon surface. After 3 min (Fig. 1.1d), a small, dark spot appears at the center of the laser-illuminated area, indicating that the temperature of the silicon surface has reached the melting point. However, no Raman peak is observed before or at the beginning of the silicon melting (Fig. 1.1g). As the melting continues, the diameter of the dark spot increases and Raman signals start appearing. After 5 min, the Raman spectrum shows an I_D/I_G ratio of around 0.076, indicating high-quality graphene with few defects. In addition, the I_D/I_{2D} ratio and the full width at half-maximum (FWHM)$_{2D}$ was found to be about 1.01 and 38 cm^{-1} respectively, showing the presence of multilayer (bi- or trilayer) graphene [45, 46]. After 15 min of irradiation, the I_D/I_{2G} ratio increased significantly, to 2.67, indicating that the graphene has no less than four layers.

Direct Growth of Graphene on Si Substrates | 5

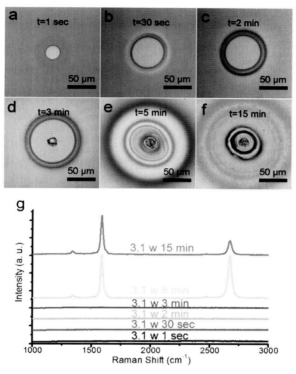

Figure 1.1 (a–f) Optical images of laser-irradiated areas. The laser illumination time is 1 s, 30 s, 2 min, 3 min, 5 min, and 15 min, respectively. (g) The corresponding Raman spectra recorded from the center of the laser-irradiated area in (a–f). Reprinted with permission from Ref. [22]. Copyright 2012, AIP Publishing LLC.

1.2.2 Carbon Ion Implantation

Carbon ion implantation is an alternative approach for obtaining graphene directly on a Si substrate. In this method, a 200–300 nm thick Ni film is first deposited on a SiO_2/Si wafer, followed by carbon ion implantation at an energy in the range of 5–80 keV, with a dose of $\sim 10^{15}$ ions/cm^2 [23, 41, 42, 47, 48]. After that, the samples are annealed at a temperature range of 600–1000°C. In most cases, the graphene layers are found over the Ni surface [23, 42, 47, 48], which still need to transfer to the SiO_2/Si substrate using the standard PMMA transfer method [49]. Kim et al. reported that by annealing the sample at 900°C, graphene layers are obtained both above and beneath the Ni film (shown in Fig. 1.2b) [41]. Figure 1.2a,d shows the corresponding Raman spectra on both surfaces (top and bottom) of the Ni film. The Raman I_D/I_G ratio of top and bottom graphene was found to be about 0.015 and 0.18, respectively, indicating the better-quality graphene grown over Ni the layer. The cross-sectional transmission electron microscopy (TEM) image (Fig. 1.2c) of the sample revealed the presence of multilayer graphene on top of the Ni film. Finally, the samples were exposed to O_2 plasma and subsequently immersed in an $FeCl_3$ solution to remove the top graphene layers and the Ni film underneath, respectively.

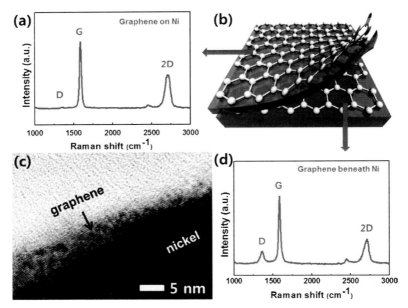

Figure 1.2 (a) Raman spectrum from graphene synthesized on top of Ni. Note that the Raman spectrum was obtained after graphene was transferred to the SiO_2/Si substrate. (b) Schematic of the graphene synthesized on both sides of Ni by ion implantation. (c) Cross-sectional TEM image showing graphene synthesized on Ni. (d) Raman spectrum from graphene synthesized at the bottom of Ni. Note that the Raman spectrum was obtained after removal of top graphene and Ni. Reprinted with permission from Ref. [41]. Copyright 2015, AIP Publishing LLC.

1.2.3 MBE Growth

MBE growth of graphene is an additional method emerging, together with carbon ion implantation [43, 44]. Herein, the MBE carbon source is present either in gaseous or solid form. In the gas-source-based MBE, ethanol gas is supplied (0.2 to 2.0 sccm) to the system with a cracking unit consisting of a tungsten (W) filament [43, 50]. The cracked ethanol produces graphitic material (graphene/amorphous carbon) on the Si(111) surface at 600°C. Maeda et al. reported that the graphitic material deposition rate decreases with an increasing gas supply rate [43]. They further found that the previously deposited graphitic material can be removed from the substrate surface by further supplying cracked ethanol (1.9 sccm) to the sample. In brief, deposition and etching occur simultaneously under the supply of cracked ethanol [43]. On the other hand, Hackley et al. reported a solid-source-based MBE graphene growth. The process starts with the deposition of a Si buffer layer of ~20 nm on the Si(111) substrate at 560°C [44]. This step provides a smooth starting surface for the subsequent graphene growth. An amorphous carbon buffer layer is than deposited via the MBE method at a relatively lower temperature. This step prevents the formation of SiC precipitates. After that, the samples were annealed at 560°C–830°C in ultrahigh vacuum. Figure 1.3 compares the C 1s photoelectron spectra of the annealed samples with a highly oriented pyrolytic graphite (HOPG) sample [44]. At 600°C, a small SiC shoulder is visible with a dominant C 1s peak. The C 1s peak further sharpens at 660°C. At 700°C, nearly all of the carbon atoms are found to be bonded to Si atoms, an indication of SiC formation. After growth at 830°C, the C 1s spectra finally indicate the formation of graphitic carbon (g-C) on the Si(111) substrate.

1.3 Thermal Decomposition of Bulk SiC

Thermal decomposition of SiC has been intensively studied lately as a promising route for obtaining highly reproducible and homogeneous large-area graphene for electronic applications [51]. The main advantage of the thermal decomposition of SiC over other conventional techniques is that the graphene layers can be directly obtained on a commercially available semiconducting or semi-insulating substrate, so no transfer is required before processing electronic devices [51–55]. Furthermore, graphene on SiC is favorable since SiC is a well-established substrate for high-frequency electronics, radiation hard devices, and light-emitting devices.

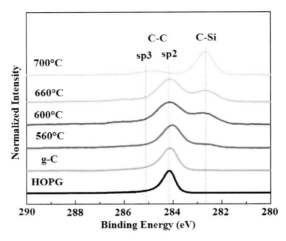

Figure 1.3 Normalized XPS spectra of the C 1s photoelectron region for various samples grown at the indicated temperatures. The position of the carbon peaks from sp^2 is 284.1 eV and sp^3 is 285.1 eV. Carbon bound to silicon is shown at 282.6 eV. The pure sp^2 nature of the g-C sample is shown along with that of HOPG, which is used as an sp^2 reference. Reprinted with permission from Ref. [44]. Copyright 2009, AIP Publishing LLC.

Although the epitaxial growth of graphene on SiC has been known for long time [56, 57], its electronic properties remained unknown until the early twenty-first century. In 2004, de Heer et al. at the Georgia Institute of Technology performed the first transport measurements on multilayer epitaxial graphene grown via thermal decomposition of SiC in ultrahigh vacuum [10, 51, 58]. They revealed the Dirac nature of the charge carriers and found the mobility values exceeding $\mu = 1,100$ cm^2/V·s in graphene on SiC. Higher mobility values were further obtained by improving the quality of graphene using the confinement-controlled sublimation (CCS) method [51, 58]. In 2010, they fabricated an array of 10,000 top-gated transistors on a 0.24 cm^2 chip [59], which is believed to be the highest density reported so far in graphene devices. For further details on their work, see Chapter 6.

Even though the thermal decomposition of SiC in high/ultrahigh vacuum appears promising for large-scale production of graphene-based devices, graphene grown via this technique consists of small grains (30–200 nm) with varying thicknesses [60, 61]. Furthermore, the quality of graphene produced in high/ultrahigh vacuum is poor due to the high sublimation rates at relatively low temperatures. To address these issues, Emtsev et al. in 2009 demonstrated a novel approach to obtaining morphologically superior graphene layers on the SiC surface [52]. This method involves annealing the SiC samples at a high temperature (>1650°C) in an argon (Ar) environment. The graphene domains obtained here are much larger in size (3 μm × 50 μm). Later, several researchers demonstrated a wide range of domain sizes, with reports of sizes as large as 50 μm × 50 μm [62, 63].

Epitaxial graphene can be grown on either of the two polar faces of a SiC crystal. The growth rate and the electronic properties are found to be dependent on the specific polar SiC crystal face. The graphene grows much faster and thicker on the $SiC(000\bar{1})$ than on SiC(0001) [64, 65]. Tedesco et al. reported a very high carrier mobility, of ~150,000 $cm^2/V\cdot s$, when graphene is grown on SiC $(000\bar{1})$ [66]. This value can reach up to ~250,000 $cm^2/V\cdot s$ for the low-temperature measurement in a magnetic field below 50 mT [67, 68]. On the contrary, a low Hall mobility value of ~5800 $cm^2/V\cdot s$ is reported for graphene on SiC(0001) [66].

Typically, single-crystalline monolayer (ML) graphene is obtained on SiC(0001) with good reproducibility [65, 69, 70]. Note that the graphene is not directly grown on top of the SiC(0001) surface but rather on a complex $(6\sqrt{3}\times6\sqrt{3})R30°$ nonconducting, carbon-rich interfacial layer [69, 71–74], which is partially covalently bonded to the underlying SiC substrate. This interfacial layer acts as an electronic buffer layer between graphene and the SiC substrate and provides a template for subsequent graphene growth [73]. A detailed discussion on decoupling this buffer layer by intercalation is further presented in Chapter 5.

In contrast, a polycrystalline graphene film is obtained on SiC $(000\bar{1})$ [56, 75–77]. Multilayer graphene grown on $SiC(000\bar{1})$ is rotationally disordered and defective [72, 78–80]. The graphene layers are found to be ordered in a particular way, with alternating 0° and 30° rotations relative to the substrate [73]. Due to this type of non-Bernal stacking, the symmetry between the atoms in the unit cell is not broken in multilayers. As a result, each graphene layer possesses the electronic structure of an isolated graphene sheet.

1.4 Graphene on Silicon through Heteroepitaxial 3C-SiC

Although very high-quality graphene has been achieved on bulk SiC through the thermal decomposition technique, the use of SiC wafers leads to limitations in terms of wafer sizes, wafer cost, and availability of micromachining processes.

Direct growth of graphene on 3C-SiC on silicon (Si) substrates would overcome such limitations [81, 82]. There are two major advantages associated with using Si as a substrate. First, Si wafers are orders of magnitude less expensive than SiC and available in large sizes of up to 12 inches. Second, using Si as a substrate provides easy access to the well-established Si-based integrated circuit technology and infrastructure [83–85].

Among the various methods investigated to grow graphene directly on 3C-SiC on Si, two appear as the most promising: (i) thermal decomposition of 3C-SiC on a Si substrate and (ii) metal-mediated graphene growth. We will discuss these methods in the following sections.

1.4.1 Thermal Decomposition of 3C-SiC on Si

Thermal decomposition is the most widely investigated technique for obtaining epitaxial graphene on 3C-SiC on Si. Several research groups have elegantly presented this technique in their respective reports, with slightly different process conditions [53, 54, 86–91]. The first formal report on the epitaxial growth of graphene on 3C-SiC(111)/Si(110) was published by Miyamoto et al. in 2009 [89]. Their graphene growth process consists of two steps: (1) growing a 3C-SiC film on a Si substrate via a gas-source MBE (GSMBE) using monomethylsilane (MMS, 99.999%) as a single source and (2) annealing the samples in ultrahigh vacuum at ~1300°C for 30 min to obtain epitaxial graphene. Figure 1.4a shows the C 1s core-level spectrum of the graphene grown on 3C-SiC on Si via this technique, indicating the formation of a well-ordered 2D network of sp^2-bonded carbon atoms [90], also suggested by Aristov et al. (see Chapter 2) [92]. They named this epitaxial growth method as graphene on silicon (GOS). Although their results were very promising, a detailed investigation later clarified that the graphene actually grew on 3C-SiC(110)/Si(110) instead of 3C-SiC(111)/Si(110) [90]. In successive years, their group produced graphene on 3C-SiC(100)/Si(100) and 3C-SiC(111)/Si(111) substrates as well [87]. Recently, they succeeded in fabricating top-gate and back-gate field effect transistors using GOS as a channel (GOSFET) [90]. The top-gate GOSFETs appeared superior for practical devices mainly in high-frequency applications because the influence of the substrate to the device performance can be eliminated. A schematic of the top-gate GOSFET is shown in Fig. 1.4b [90]. The high-resolution transmission electron micrograph indicates a multilayer graphene grown on the 3C-SiC surface. The transfer characteristics of this top-gate GOSFET revealed an ambipolar behavior with a minimum conductance at the gate bias voltage of 3.8 V.

Among other research groups, Ouerghi's group from the Centre National de la Recherche Scientifique (CNRS), France, is a major contributor in developing this high-vacuum sublimation process and published three consecutive papers in 2010, explaining the graphene growth mechanism [81, 86, 93]. They investigated the structural and electronic properties of epitaxial graphene on a 3C-SiC /Si(111) substrate using scanning tunneling microscopy (STM), low-energy electron diffraction (LEED), scanning transmission electron microscopy (STEM), and synchrotron angle-resolved photoemission spectroscopy (ARPES). Their result shows that graphene has remarkable continuity on terraces and step edges, suggesting the possibility of large-scale graphene production. In another article, they studied graphene growth on a 3C-SiC film deposited on on-axis and off-axis Si(100) substrates [93, 94]. As we know, the 3C-SiC heteroepitaxial films grown on Si(100) substrates possess a high density of defects, such as antiphase boundaries (APBs) [95, 96]. These APBs can transfer to the graphene layers with dissimilar thickness, which deteriorate the intrinsic properties of ideal graphene. Ouerghi et al. reported that the 3C-SiC films grown on off-axis Si(100) substrates can be used to eliminate these APBs, and single-domain epitaxial graphene films are obtained on the surface [94]. For more details on Ouerghi's work, see Chapter 4.

1.4.2 Metal-Mediated Graphene Growth

The thermal decomposition approach appears to have two major limitations in the case of silicon carbide on silicon. First, the quality of the graphene film produced in high/ultrahigh vacuum is limited due to the difficulty in controlling sublimation rates at relatively low (900°C–1300°C) temperatures [53, 89, 97]. A large D to G band Raman intensity ratio (I_D/I_G) of ~1 was reported, which is considerably high in comparison to the exfoliated graphene [53, 89]. Second, the thermal decomposition technique is commonly limited to the use of the 3C-SiC/Si(111) substrate [53, 81, 90], whereas Si(100) is the most preferred substrate in the microelectronics industry due to the lower atom density at the surface [98], which results in high carrier mobility. Thus, an alternative method is required to obtain high-quality graphene on both (100)- and (111)-oriented Si substrates.

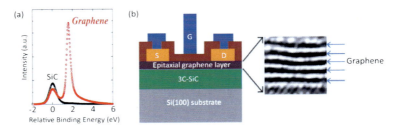

Figure 1.4 (a) Comparison of C 1s core-level spectra of epitaxial 3C-SiC (110) before (black) and after (red) thermal annealing at 1300°C for 30 min. After graphitization, the sp^2 peak is strongly enhanced. (b) Schematic cross section of a GOSFET (left) and TEM image of an epitaxial graphene layer (right). Copyright © 2013 IEEE. Reprinted, with permission, from Ref. [90].

To address these issues, a few groups have investigated an alternative catalyst-based approach for obtaining graphene on the 3C-SiC on Si [99–102]. This method involves depositing a metal layer such as nickel or cobalt on the 3C-SiC surface and subsequently annealing the sample at temperatures ranging from 750°C to 1200°C, which is much lower than that for the thermal decomposition process. During the annealing process SiC reacts with metal, forming a metal silicide and releasing atomic carbon into the system [103]. This released atomic carbon precipitate upon cooling rearranges into single- to few-layer graphene. In most cases the graphene was found to grow on the metal surface, still needing to be transferred onto a semiconductor or an insulating surface to obtain a functional device, which hinders its utility for large-scale device fabrication [99, 101]. Although the earlier attempts at nickel-mediated graphitization from amorphous or crystalline SiC films on silicon had shown some promise, improving the defect density and uniformity of the graphene remained challenging.

In response to these limitations, we have recently demonstrated a novel alloy-mediated catalytic approach to growing high-quality, highly uniform bilayer graphene on 3C-SiC on both Si(100) and Si(111) substrates [104, 105]. The graphitization process consists of the following steps: (1) deposition of double layer of nickel (Ni) and copper (Cu) onto the 3C-SiC surface; (2) annealing of samples at a mild temperature (900°C–1100°C) for 1 h; and (3) removal of the metal/metal-silicide layer by immersing samples in a wet chemical etch solution. The obtained bilayer graphene covers uniformly a 2-inch silicon wafer with an average Raman I_D/I_G band ratio as low as 0.2 to 0.3, indicative of a lowly defective material [104]. Note that this I_D/I_G ratio is consistent over large surfaces and considerably small compared to the previously reported graphene on 3C-SiC on Si [53, 90, 93].

Due to its extraordinary mechanical and electrical properties [7, 106], a graphene coating is expected to increase the sensitivity of commercially available microelectromechanical systems (MEMS) [107]. However, the low adhesion of transferred graphene is a major drawback for device fabrication and reliability. The alloy-mediated graphene has indicated an adhesion energy to the underlying 3C-SiC film [104] nearly 1 order of magnitude higher than that of graphene layers transferred onto a SiO_2 layer on Si [20]. Additionally, wafer-scale fabrication of graphitized SiC microstructures (bridges and cantilever) on a silicon substrate through a selective, self-aligned, transfer-free catalytic process was recently demonstrated. Figure 1.5a illustrates the complete fabrication process [105]. First, the microstructures are patterned on the 3C-SiC surface via a standard photolithography technique. In this, the SiC/Si wafer is coated with photoresist, and the structures are defined. Unprotected SiC areas are selectively removed via HCl plasma etching, and the remaining photoresist is removed using O_2 plasma. The bimetal (Ni and Cu) catalyst layer is deposited on the entire wafer. After that, the samples are annealed at a moderate temperature (1100°C) in a Carbolite HT furnace for 1 h to produce a graphene and a metal/metal-silicide layer on the patterned 3C-SiC surface. The metal/metal-silicide layer is then removed by immersing the samples in a wet chemical etch solution. Finally, a XeF$_2$ isotropic silicon etching is performed to release the graphitized 3C-SiC structures. Figure 1.5b,c shows the scanning electron microscopy (SEM) images of graphitized 3C-SiC microstructures including bridges and cantilevers fabricated via this technique [105]. Finally, we have recently demonstrated that the metal-mediated graphene approach using only nickel can also be optimized to obtain a highly rugged electrode material for on-chip supercapacitors [108].

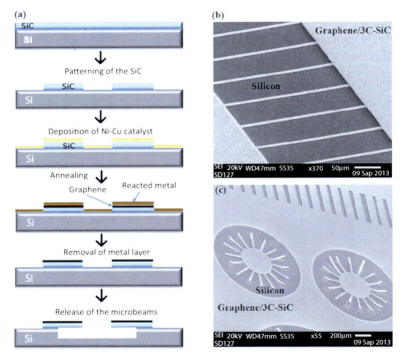

Figure 1.5 (a) Sequential steps for the wafer-level fabrication of graphitized silicon carbide microbeams on a silicon substrate. Once the SiC is patterned, the few-layer graphene is grown selectively on the SiC structures via metal-mediated graphitization. The reacted metal layer is subsequently removed, and the structures are released from the substrate to form suspended beams. SEM micrographs of released graphene/3C-SiC microstructures including (b) bridges and (c) cantilevers. Adapted from Ref. [105]. Copyright © IOP Publishing. Reproduced with permission. All rights reserved.

Therefore, we suggest that although the graphene quality may not match that of graphene grown on silicon carbide wafers, metal-mediated graphene growth from heteroepitaxial silicon carbide offers large flexibility and opens the door for large-scale fabrication of graphene nanostructures for a large range of electronic, photonic, optomechanical, energy storage, and sensing applications on a silicon platform.

1.5 Conclusions

For commercialization of graphene-based devices, fabrication needs to be simple, reproducible, and compatible with existing semiconductor technology. This implies that many of the existing techniques for graphene synthesis will have to be change drastically. Hence, the direct growth of high-quality graphene onto commonly used materials in the semiconductor industry is highly desirable. This chapter has provided an overview of direct graphene growth on semiconducting surfaces, particularly on SiC and silicon, and their properties and technological relevance.

Graphene on SiC is most promising and advanced at the moment from the perspective of electronic device fabrication, thanks to several novel techniques developed for obtaining large-area, high-quality graphene on it. Although the SiC substrate has a potential for mass production of graphene, there are several technological limitations. SiC wafers are limited in size ($\sim$3 inches), very expensive, and difficult to micromachine. Therefore, an alternative appropriate substrate is required.

Since silicon is the most widely used semiconductor in the electronic industry, the direct integration of graphene on silicon seems beneficial for future device applications. However, the currently available direct graphene growth techniques on silicon substrate produce graphene flakes, dots, and graphitic (amorphous) carbon, rather than a continuous film. An alternative method using a SiC heteroepitaxial film on a large-area silicon substrate (up to 12 inches) has been proposed for high-quality graphene growth. A SiC film on a silicon substrate offers several advantages, such as compatibility with established silicon fabrication technologies, less fabrication costs, and availability of a large-area substrate.

Moreover, a recently developed catalyst alloy-mediated graphene approach on heteroepitaxial SiC opens the possibility for direct and selective growth of high-quality graphene on silicon, leading to straightforward patterning and micromachining capabilities.

Acknowledgments

The authors would like to acknowledge funding support from the Air Force Office of Scientific Research, Office of Naval Research Global, and Army Research Development and Engineering Command (RDECOM) through the grant AFOSR/AOARD 15IOA053. F. Iacopi is the recipient of an Australian Research Council Future Fellowship (FT120100445). N. Motta acknowledges the financial support of the Australian Research Council (ARC) through the Discovery Project DP130102120 and the LIEF grant LE100100146.

References

1. Novoselov, K. S., Geim, A. K., Morozov, S. V., Jiang, D., Katsnelson, M. I., Grigorieva, I. V., Dubonos, S. V., and Firsov, A. A., Two-dimensional gas of massless Dirac fermions in graphene, *Nature*, **438**, 197–200 (2005).

2. Gusynin, V. P., and Sharapov, S. G., Unconventional integer quantum Hall effect in graphene, *Phys. Rev. Lett.*, **95**, 146801 (2005).

3. Zhang, Y., Tan, Y.-W., Stormer, H. L., and Kim, P., Experimental observation of the quantum Hall effect and Berry's phase in graphene, *Nature*, **438**, 201–204 (2005).

4. Morozov, S. V., Novoselov, K. S., Katsnelson, M. I., Schedin, F., Ponomarenko, L. A., Jiang, D., and Geim, A. K., Strong suppression of weak localization in graphene, *Phys. Rev. Lett.*, **97**, 016801 (2006).

5. Geim, A. K., and Novoselov, K. S., The rise of graphene, *Nat. Mater.*, **6**, 183–191 (2007).

6. Bolotin, K. I., Sikes, K., Jiang, Z., Klima, M., Fudenberg, G., Hone, J., Kim, P., and Stormer, H., Ultrahigh electron mobility in suspended graphene, *Solid State Commun.*, **146**, 351–355 (2008).

7. Novoselov, K. S., Geim, A. K., Morozov, S. V., Jiang, D., Zhang, Y., Dubonos, S. V., Grigorieva, I. V., and Firsov, A. A., Electric field effect in atomically thin carbon films, *Science*, **306**, 666–669 (2004).

8. Jiao, L., Zhang, L., Wang, X., Diankov, G., and Dai, H., Narrow graphene nanoribbons from carbon nanotubes, *Nature*, **458**, 877–880 (2009).

9. Choucair, M., Thordarson, P., and Stride, J. A., Gram-scale production of graphene based on solvothermal synthesis and sonication, *Nat. Nanotechnol.*, **4**, 30–33 (2008).

10. Berger, C., Song, Z., Li, T., Li, X., Ogbazghi, A. Y., Feng, R., Dai, Z., Marchenkov, A. N., Conrad, E. H., and First, P. N., Conversion of carbon dioxide to few-layer graphene, *J. Phys. Chem. B*, **108**, 19912–19916 (2004).

11. Chakrabarti, A., Lu, J., Skrabutenas, J. C., Xu, T., Xiao, Z., Maguire, J. A., and Hosmane, N. S., On the role of vapor trapping for chemical vapor deposition (CVD) grown graphene over copper, *J. Mater. Chem.*, **21**, 9491–9493 (2011).

12. Bae, S., Kim, H., Lee, Y., Xu, X., Park, J.-S., Zheng, Y., Balakrishnan, J., Lei, T., Ri Kim, H., Song, Y. I., Kim, Y.-J., Kim, K. S., Ozyilmaz, B., Ahn, J.-H., Hong, B. H., and Iijima, S., Large-area synthesis of high-quality and uniform graphene films on copper foils, *Nat. Nanotechnol.*, **5**, 574–578 (2010).

13. Rümmeli, M. H., Gorantla, S., Bachmatiuk, A., Phieler, J., Geißler, N., Ibrahim, I., Pang, J., and Eckert, J., Ultrathin epitaxial graphite: 2D electron gas properties and a route toward graphene-based nanoelectronics, *Chem. Mater.*, **25**, 4861–4866 (2013).

14. Rümmeli, M. H., Bachmatiuk, A., Scott, A., Börrnert, F., Warner, J. H., Hoffman, V., Lin, J.-H., Cuniberti, G., and Büchner, B., Roll-to-roll production of 30-inch graphene films for transparent electrodes, *ACS Nano*, **4**, 4206–4210 (2010).

15. Li, X., Cai, W., An, J., Kim, S., Nah, J., Yang, D., Piner, R., Velamakanni, A., Jung, I., and Tutuc, E., Direct low-temperature nanographene CVD synthesis over a dielectric insulator, *Science*, **324**, 1312–1314 (2009).

16. Du, X., Skachko, I., Barker, A., and Andrei, E. Y., Approaching ballistic transport in suspended graphene, *Nat. Nanotechnol.*, **3**, 491–495 (2008).

17. Ni, G.-X., Zheng, Y., Bae, S., Kim, H. R., Pachoud, A., Kim, Y. S., Tan, C.-L., Im, D., Ahn, J.-H., Hong, B. H., and Özyilmaz, B., Quasi-periodic nanoripples in graphene grown by chemical vapor deposition and its impact on charge transport, *ACS Nano*, **6**, 1158–1164 (2012).

18. Zang, J., Ryu, S., Pugno, N., Wang, Q., Tu, Q., Buehler, M. J., and Zhao, X., Multifunctionality and control of the crumpling and unfolding of large-area graphene, *Nat. Mater.*, **12**, 321–325 (2013).

19. Lin, Y.-M., Valdes-Garcia, A., Han, S.-J., Farmer, D. B., Meric, I., Sun, Y., Wu, Y., Dimitrakopoulos, C., Grill, A., and Avouris, P., Wafer-scale graphene integrated circuit, *Science*, **332**, 1294–1297 (2011).

20. Koenig, S. P., Boddeti, N. G., Dunn, M. L., and Bunch, J. S., Ultrastrong adhesion of graphene membranes, *Nat. Nanotechnol.*, **6**, 543–546 (2011).

21. Yoon, T., Shin, W. C., Kim, T. Y., Mun, J. H., Kim, T.-S., and Cho, B. J., Direct measurement of adhesion energy of monolayer graphene as-grown on copper and its application to renewable transfer process, *Nano Lett.*, **12**, 1448–1452 (2012).

22. Wei, D., and Xu, X., Laser direct growth of graphene on silicon substrate, *Appl. Phys. Lett.*, **100**, 023110 (2012).

23. Mun, J. H., Lim, S. K., and Cho, B. J., Local growth of graphene by ion implantation of carbon in a nickel thin film followed by rapid thermal annealing, *J. Electrochem. Soc.*, **159**, G89–G92 (2012).

24. Michon, A., Tiberj, A., Vézian, S., Roudon, E., Lefebvre, D., Portail, M., Zielinski, M., Chassagne, T., Camassel, J., and Cordier, Y., Graphene growth on AlN templates on silicon using propane-hydrogen chemical vapor deposition, *Appl. Phys. Lett.*, **104**, 071912 (2014).

25. Chen, J., Wen, Y., Guo, Y., Wu, B., Huang, L., Xue, Y., Geng, D., Wang, D., Yu, G., and Liu, Y., Oxygen-aided synthesis of polycrystalline graphene on silicon dioxide substrates, *J. Am. Chem. Soc.*, **133**, 17548–17551 (2011).

26. Wang, G., Zhang, M., Zhu, Y., Ding, G., Jiang, D., Guo, Q., Liu, S., Xie, X., Chu, P. K., and Di, Z., Direct growth of graphene film on germanium substrate, *Sci. Rep.*, **3** (2013).

27. Tang, S., Ding, G., Xie, X., Chen, J., Wang, C., Ding, X., Huang, F., Lu, W., and Jiang, M., Nucleation and growth of single crystal graphene on hexagonal boron nitride, *Carbon*, **50**, 329–331 (2012).

28. Hwang, J., Kim, M., Campbell, D., Alsalman, H. A., Kwak, J. Y., Shivaraman, S., Woll, A. R., Singh, A. K., Hennig, R. G., Gorantla, S., Rümmeli, M. H., and Spencer, M. G., van der Waals epitaxial growth of graphene on sapphire by chemical vapor deposition without a metal catalyst, *ACS Nano*, **7**, 385–395 (2013).

29. Ismach, A., Druzgalski, C., Penwell, S., Schwartzberg, A., Zheng, M., Javey, A., Bokor, J., and Zhang, Y., Direct chemical vapor deposition of graphene on dielectric surfaces, *Nano Lett.*, **10**, 1542–1548 (2010).

30. Pasternak, I., Wesolowski, M., Jozwik, I., Lukosius, M., Lupina, G., Dabrowski, P., Baranowski, J. M., and Strupinski, W., Graphene growth on Ge (100)/Si (100) substrates by CVD method, *Sci. Rep.*, **6**, 21773 (2016).

31. Lin, Y.-M., Dimitrakopoulos, C., Jenkins, K. A., Farmer, D. B., Chiu, H.-Y., Grill, A., and Avouris, P., 100-GHz transistors from wafer-scale epitaxial graphene, *Science*, **327**, 662–662 (2010).

32. Xia, F., Farmer, D. B., Lin, Y.-M., and Avouris, P., Graphene field-effect transistors with high on/off current ratio and large transport band gap at room temperature, *Nano Lett.*, **10**, 715–718 (2010).

33. Ichinokura, S., Sugawara, K., Takayama, A., Takahashi, T., and Hasegawa, S., Superconducting calcium-intercalated bilayer graphene, *ACS Nano*, **10**, 2761–2765 (2016).

34. Bi, H., Sun, S., Huang, F., Xie, X., and Jiang, M., Direct growth of few-layer graphene films on SiO_2 substrates and their photovoltaic applications, *J. Mater. Chem.*, **22**, 411–416 (2012).

35. Kim, K., Choi, J.-Y., Kim, T., Cho, S.-H., and Chung, H.-J., A role for graphene in silicon-based semiconductor devices, *Nature*, **479**, 338–344 (2011).

36. Bresnehan, M. S., Hollander, M. J., Wetherington, M., Wang, K., Miyagi, T., Pastir, G., Snyder, D. W., Gengler, J. J., Voevodin, A. A., Mitchel, W. C., and Robinson, J. A., Prospects of direct growth boron nitride films as substrates for graphene electronics, *J. Mater. Res.*, **29**, 459–471 (2014).

37. Lemme, M. C., Echtermeyer, T. J., Baus, M., and Kurz, H., Graphene field-effect device, *IEEE Electron Device Lett.*, **28**, 282–284 (2007).

38. Lin, Y.-M., Jenkins, K. A., Valdes-Garcia, A., Small, J. P., Farmer, D. B., and Avouris, P., Operation of graphene transistors at gigahertz frequencies, *Nano Lett.*, **9**, 422–426 (2009).

39. Farmer, D. B., Chiu, H.-Y., Lin, Y.-M., Jenkins, K. A., Xia, F., and Avouris, P., Utilization of a buffered dielectric to achieve high field-effect carrier mobility in graphene transistors, *Nano Lett.*, **9**, 4474–4478 (2009).

40. Liao, L., Bai, J., Cheng, R., Lin, Y.-C., Jiang, S., Huang, Y., and Duan, X., Top-gated graphene nanoribbon transistors with ultrathin high-k dielectrics, *Nano Lett.*, **10**, 1917–1921 (2010).

41. Kim, J., Lee, G., and Kim, J., Wafer-scale synthesis of multi-layer graphene by high-temperature carbon ion implantation, *Appl. Phys. Lett.*, **107**, 033104 (2015).

42. Gutierrez, G., Le Normand, F., Muller, D., Aweke, F., Speisser, C., Antoni, F., Le Gall, Y., Lee, C. S., and Cojocaru, C. S., Multi-layer graphene obtained by high temperature carbon implantation into nickel films, *Carbon*, **66**, 1–10 (2014).

43. Maeda, F., and Hibino, H., Study of graphene growth by gas-source molecular beam epitaxy using cracked ethanol: influence of gas flow

rate on graphitic material deposition, *Jpn. J. Appl. Phys.*, **50**, 06GE12 (2011).

44. Hackley, J., Ali, D., DiPasquale, J., Demaree, J. D., and Richardson, C. J. K., Graphitic carbon growth on Si(111) using solid source molecular beam epitaxy, *Appl. Phys. Lett.*, **95**, 133114 (2009).

45. Li, X., Cai, W., An, J., Kim, S., Nah, J., Yang, D., Piner, R., Velamakanni, A., Jung, I., Tutuc, E., Banerjee, S. K., Colombo, L., and Ruoff, R. S., Large-area synthesis of high-quality and uniform graphene films on copper foils, *Science*, **324**, 1312–1314 (2009).

46. Ferrari, A. C., and Basko, D. M., Raman spectroscopy as a versatile tool for studying the properties of graphene, *Nat. Nanotechnol.*, **8**, 235–246 (2013).

47. Laurent, B., Zhanbing, H., Chang Seok, L., Jean-Luc, M., Costel Sorin, C., Anne-Françoise, G.-L., Young Hee, L., and Didier, P., Synthesis of few-layered graphene by ion implantation of carbon in nickel thin films, *Nanotechnology*, **22**, 085601 (2011).

48. Garaj, S., Hubbard, W., and Golovchenko, J. A., Graphene synthesis by ion implantation, *Appl. Phys. Lett.*, **97**, 183103 (2010).

49. Reina, A., Jia, X., Ho, J., Nezich, D., Son, H., Bulovic, V., Dresselhaus, M. S., and Kong, J., Large area, few-layer graphene films on arbitrary substrates by chemical vapor deposition, *Nano Lett.*, **9**, 30–35 (2008).

50. Maeda, F., and Hibino, H., Growth of few-layer graphene by gas-source molecular beam epitaxy using cracked ethanol, *Phys. Status Solidi B*, **247**, 916–920 (2010).

51. de Heer, W. A., Berger, C., Ruan, M., Sprinkle, M., Li, X., Hu, Y., Zhang, B., Hankinson, J., and Conrad, E., Large area and structured epitaxial graphene produced by confinement controlled sublimation of silicon carbide, *Proc. Natl. Acad. Sci. U S A*, **108**, 16900–16905 (2011).

52. Emtsev, K. V., Bostwick, A., Horn, K., Jobst, J., Kellogg, G. L., Ley, L., McChesney, J. L., Ohta, T., Reshanov, S. A., and Röhrl, J., Towards wafer-size graphene layers by atmospheric pressure graphitization of silicon carbide, *Nat. Mater.*, **8**, 203–207 (2009).

53. Gupta, B., Notarianni, M., Mishra, N., Shafiei, M., Iacopi, F., and Motta, N., Evolution of epitaxial graphene layers on 3C SiC/Si (111) as a function of annealing temperature in UHV, *Carbon*, **68**, 563–572 (2014).

54. Gupta, B., Placidi, E., Hogan, C., Mishra, N., Iacopi, F., and Motta, N., The transition from 3C SiC (111) to graphene captured by ultra high vacuum scanning tunneling microscopy, *Carbon*, **91**, 378–385 (2015).

55. Seyller, T., Bostwick, A., Emtsev, K. V., Horn, K., Ley, L., McChesney, J. L., Ohta, T., Riley, J. D., Rotenberg, E., and Speck, F., Epitaxial graphene: a new material, *Phys. Status Solidi B*, **245**, 1436–1446 (2008).

56. Van Bommel, A. J., Crombeen, J. E., and Van Tooren, A., LEED and Auger electron observations of the SiC(0001) surface, *Surf. Sci.*, **48**, 463–472 (1975).

57. Badami, D., X-ray studies of graphite formed by decomposing silicon carbide, *Carbon*, **3**, 53–57 (1965).

58. Berger, C., Song, Z., Li, X., Wu, X., Brown, N., Naud, C., Mayou, D., Li, T., Hass, J., and Marchenkov, A. N., Electronic confinement and coherence in patterned epitaxial graphene, *Science*, **312**, 1191–1196 (2006).

59. Sprinkle, M., Ruan, M., Hu, Y., Hankinson, J., Rubio-Roy, M., Zhang, B., Wu, X., Berger, C., and de Heer, W. A., Scalable templated growth of graphene nanoribbons on SiC, *Nat. Nanotechnol.*, **5**, 727–731 (2010).

60. Hass, J., Feng, R., Li, T., Li, X., Zong, Z., de Heer, W. A., First, P. N., Conrad, E. H., Jeffrey, C. A., and Berger, C., Highly ordered graphene for two dimensional electronics, *Appl. Phys. Lett.*, **89**, 143106 (2006).

61. Hibino, H., Kageshima, H., Maeda, F., Nagase, M., Kobayashi, Y., and Yamaguchi, H., Microscopic thickness determination of thin graphite films formed on SiC from quantized oscillation in reflectivity of low-energy electrons, *Phys. Rev. B*, **77**, 075413 (2008).

62. Yazdi, G. R., Vasiliauskas, R., Iakimov, T., Zakharov, A., Syväjärvi, M., and Yakimova, R., Growth of large area monolayer graphene on 3C-SiC and a comparison with other SiC polytypes, *Carbon*, **57**, 477–484 (2013).

63. Tedesco, J. L., Jernigan, G. G., Culbertson, J. C., Hite, J. K., Yang, Y., Daniels, K. M., Myers-Ward, R. L., Eddy, C. R., Robinson, J. A., Trumbull, K. A., Wetherington, M. T., Campbell, P. M., and Gaskill, D. K., Morphology characterization of argon-mediated epitaxial graphene on C-face SiC, *Appl. Phys. Lett.*, **96**, 222103 (2010).

64. Jernigan, G. G., VanMil, B. L., Tedesco, J. L., Tischler, J. G., Glaser, E. R., Davidson, A., Campbell, P. M., and Gaskill, D. K., Comparison of epitaxial graphene on Si-face and C-face 4H SiC formed by ultrahigh vacuum and RF furnace production, *Nano Lett.*, **9**, 2605–2609 (2009).

65. Srivastava, L. N., He, G., Feenstra, R. M., and Fisher, P. J., Comparison of graphene formation on C-face and Si-face SiC {0001} surfaces, *Phys. Rev. B*, **82**, 235406 (2010).

66. Tedesco, J. L., VanMil, B. L., Myers-Ward, R. L., McCrate, J. M., Kitt, S. A., Campbell, P. M., Jernigan, G. G., Culbertson, J. C., Eddy, C., and Gaskill, D.

K., Hall effect mobility of epitaxial graphene grown on silicon carbide, *Appl. Phys. Lett.*, **95**, 122102 (2009).

67. Orlita, M., Faugeras, C., Plochocka, P., Neugebauer, P., Martinez, G., Maude, D. K., Barra, A. L., Sprinkle, M., Berger, C., de Heer, W. A., and Potemski, M., Approaching the Dirac point in high-mobility multilayer epitaxial graphene, *Phys. Rev. Lett.*, **101**, 267601 (2008).

68. Chiang, S., Enriquez, H., Oughaddou, H., Soukiassian, P., Gala, A. T., and Vizzini, S., *Graphene Epitaxied on SiC, with an Open Band Gap and Mobility Comparable to Standard Graphene with Zero Band Gap*, US Patent 20130126865, (2013)..

69. Lauffer, P., Emtsev, K. V., Graupner, R., Seyller, T., Ley, L., Reshanov, S. A., and Weber, H. B., Atomic and electronic structure of few-layer graphene on SiC(0001) studied with scanning tunneling microscopy and spectroscopy, *Phys. Rev. B*, **77**, 155426 (2008).

70. Virojanadara, C., Yakimova, R., Zakharov, A. A., and Johansson, L. I., *J. Large homogeneous mono-/bi-layer graphene on 6H–SiC(0 0 0 1) and buffer layer elimination, *Phys. D: Appl. Phys.*, **43**, 374010 (2010).

71. Varchon, F., Feng, R., Hass, J., Li, X., Nguyen, B. N., Naud, C., Mallet, P., Veuillen, J. Y., Berger, C., Conrad, E. H., and Magaud, L., Electronic structure of epitaxial graphene layers on SiC: effect of the substrate, *Phys. Rev. Lett.*, **99**, 126805 (2007).

72. Emtsev, K. V., Speck, F., Seyller, T., Ley, L., and Riley, J. D., Ambipolar doping in quasifree epitaxial graphene on SiC(0001) controlled by Ge intercalation, *Phys. Rev. B*, **77**, 155303 (2008).

73. Ruan, M., Hu, Y., Guo, Z., Dong, R., Palmer, J., Hankinson, J., Berger, C., and de Heer, W. A., Interaction, growth, and ordering of epitaxial graphene on SiC{0001} surfaces: a comparative photoelectron spectroscopy study, *MRS Bull.*, **37**, 1138–1147 (2012).

74. Emtsev, K. V., Zakharov, A. A., Coletti, C., Forti, S., and Starke, U., Epitaxial graphene on silicon carbide: introduction to structured graphene, *Phys. Rev. B*, **84**, 125423 (2011).

75. Forbeaux, I., Themlin, J.-M., Charrier, A., Thibaudau, F., and Debever, J.-M., Solid-state graphitization mechanisms of silicon carbide 6H–SiC polar faces, *Appl. Surf. Sci.*, **162**, 406–412 (2000).

76. Hite, J. K., Twigg, M. E., Tedesco, J. L., Friedman, A. L., Myers-Ward, R. L., Eddy, C. R., and Gaskill, D. K., Epitaxial graphene nucleation on C-face silicon carbide, *Nano Lett.*, **11**, 1190–1194 (2011).

77. Nyakiti, L., Wheeler, V., Garces, N., Myers-Ward, R., Eddy, C., and Gaskill, D., Enabling graphene-based technologies: toward wafer-scale production of epitaxial graphene, *MRS Bull.*, **37**, 1149–1157 (2012).

78. Hass, J., Feng, R., Millan-Otoya, J., Li, X., Sprinkle, M., First, P., de Heer, W. A., Conrad, E., and Berger, C., Structural properties of the multilayer graphene/4 H–SiC (000 1⁻) system as determined by surface x-ray diffraction, *Phys. Rev. B*, **75**, 214109 (2007).

79. Hass, J., Varchon, F., Millan-Otoya, J.-E., Sprinkle, M., Sharma, N., de Heer, W. A., Berger, C., First, P. N., Magaud, L., and Conrad, E. H., Why multilayer graphene on 4H-SiC (0001 [over]) behaves like a single sheet of graphene, *Phys. Rev. Lett.*, **100**, 125504 (2008).

80. Hass, J., de Heer, W. A., and Conrad, E., The growth and morphology of epitaxial multilayer graphene, *J. Phys.: Condens. Matter*, **20**, 323202 (2008).

81. Ouerghi, A., Marangolo, M., Belkhou, R., El Moussaoui, S., Silly, M. G., Eddrief, M., Largeau, L., Portail, M., Fain, B., and Sirotti, F., Epitaxial graphene on 3C-SiC(111) pseudosubstrate: structural and electronic properties, *Phys. Rev. B*, **82**, 125445 (2010).

82. Miyamoto, Y., Handa, H., Saito, E., Konno, A., Narita, Y., Suemitsu, M., Fukidome, H., Ito, T., Yasui, K., Nakazawa, H., and Endoh, T., Raman-scattering spectroscopy of epitaxial graphene formed on SiC film on Si substrate, *e-J. Surf. Sci. Nanotechnol.*, **7**, 107–109 (2009).

83. Kermany, A. R., Brawley, G., Mishra, N., Sheridan, E., Bowen, W. P., and Iacopi, F., Microresonators with Q-factors over a million from highly stressed epitaxial silicon carbide on silicon, *Appl. Phys. Lett.*, **104**, 081901 (2014).

84. Ranjbar Kermany, A., and Iacopi, F., Controlling the intrinsic bending of hetero-epitaxial silicon carbide micro-cantilevers, *J. Appl. Phys.*, **118**, 155304 (2015).

85. Iacopi, F., Walker, G., Wang, L., Malesys, L., Ma, S., Cunning, B. V., and Iacopi, A., Orientation-dependent stress relaxation in hetero-epitaxial 3C-SiC films, *Appl. Phys. Lett.*, **102**, 011908 (2013).

86. Ouerghi, A., Kahouli, A., Lucot, D., Portail, M., Travers, L., Gierak, J., Penuelas, J., Jegou, P., Shukla, A., Chassagne, T., and Zielinski, M., Epitaxial graphene on cubic SiC(111)/Si(111) substrate, *Appl. Phys. Lett.*, **96**, 191910 (2010).

87. Suemitsu, M., and Fukidome, H., Epitaxial graphene on silicon substrates, *J. Phys. D: Appl. Phys.*, **43**, 374012 (2010).

88. Fanton, M., Robinson, J., Weiland, B., and Moon, J., 3C-SiC films grown on Si (111) substrates as a template for graphene epitaxy, *ECS Trans.*, **19**, 131–135 (2009).

89. Suemitsu, M., Miyamoto, Y., Handa, H., and Konno, A., Time evolution of graphene growth on SiC as a function of annealing temperature, *e-J. Surf. Sci. Nanotechnol.*, **7**, 311–313 (2009).

90. Fukidome, H., Kawai, Y., Handa, H., Hibino, H., Miyashita, H., Kotsugi, M., Ohkochi, T., Jung, M., Suemitsu, T., Kinoshita, T., Otsuji, T., and Suemitsu, M., Graphene formation on a 3C-SiC(111) thin film grown on Si(110) substrate, *Proc. IEEE*, **101**, 1557–1566 (2013).

91. Zarotti, F., Gupta, B., Iacopi, F., Sgarlata, A., Tomellini, M., and Motta, N., Site-selective epitaxy of graphene on Si wafers, *Carbon*, **98**, 307–312 (2016).

92. Aristov, V. Y., Urbanik, G., Kummer, K., Vyalikh, D. V., Molodtsova, O. V., Preobrajenski, A. B., Zakharov, A. A., Hess, C., Hänke, T., Büchner, B., Vobornik, I., Fujii, J., Panaccione, G., Ossipyan, Y. A., and Knupfer, M., Graphene synthesis on cubic SiC/Si wafers. Perspectives for mass production of graphene-based electronic devices, *Nano Lett.*, **10**, 992–995 (2010).

93. Ouerghi, A., Ridene, M., Balan, A., Belkhou, R., Barbier, A., Gogneau, N., Portail, M., Michon, A., Latil, S., Jegou, P., and Shukla, A., Sharp interface in epitaxial graphene layers on 3C-SiC(100)/Si(100) wafers, *Phys. Rev. B*, **83**, 205429 (2011).

94. Ouerghi, A., Balan, A., Castelli, C., Picher, M., Belkhou, R., Eddrief, M., Silly, M., Marangolo, M., Shukla, A., and Sirotti, F., Epitaxial graphene on single domain 3C-SiC (100) thin films grown on off-axis Si (100), *Appl. Phys. Lett.*, **101**, 021603 (2012).

95. Pirouz, P., Chorey, C. M., and Powell, J. A., Antiphase boundaries in epitaxially grown β-SiC, *Appl. Phys. Lett.*, **50**, 221–223 (1987).

96. Mishra, N., Hold, L., Iacopi, A., Gupta, B., Motta, N., and Iacopi, F., Controlling the surface roughness of epitaxial SiC on silicon, *J. Appl. Phys.*, **115**, 203501 (2014).

97. Zhang, Y., Tang, T.-T., Girit, C., Hao, Z., Martin, M. C., Zettl, A., Crommie, M. F., Shen, Y. R., and Wang, F., Direct observation of a widely tunable bandgap in bilayer graphene, *Nature*, **459**, 820–823 (2009).

98. Yates, J. T., A new opportunity in silicon-based microelectronics, *Science*, **279**, 335–336 (1998).

99. Juang, Z.-Y., Wu, C.-Y., Lo, C.-W., Chen, W.-Y., Huang, C.-F., Hwang, J.-C., Chen, F.-R., Leou, K.-C., and Tsai, C.-H., Synthesis of graphene on silicon carbide substrates at low temperature, *Carbon*, **47**, 2026–2031 (2009).

100. Hofrichter, J., Szafranek, B. u. N., Otto, M., Echtermeyer, T. J., Baus, M., Majerus, A., Geringer, V., Ramsteiner, M., and Kurz, H., Wafer scale catalytic growth of graphene on nickel by solid carbon source, *Nano Lett.*, **10**, 36–42 (2009).

101. Delamoreanu, A., Rabot, C., Vallee, C., and Zenasni, A., Synthesis of graphene on silicon dioxide by a solid carbon source, *Carbon*, **66**, 48–56 (2014).

102. Li, C., Li, D., Yang, J., Zeng, X., and Yuan, W., Preparation of single-and few-layer graphene sheets using co deposition on SiC substrate, *J. Nanomater.*, **2011**, 44 (2011).

103. Julies, B. A., Knoesen, D., Pretorius, R., and Adams, D., A study of the NiSi to $NiSi_2$ transition in the Ni–Si binary system, *Thin Solid Films*, **347**, 201–207 (1999).

104. Iacopi, F., Mishra, N., Cunning, B. V., Goding, D., Dimitrijev, S., Brock, R., Dauskardt, R. H., Wood, B., and Boeckl, J., A catalytic alloy approach for graphene on epitaxial SiC on silicon wafers, *J. Mater. Res.*, **30**, 609–616 (2015).

105. Cunning, B. V., Ahmed, M., Mishra, N., Kermany, A. R., Wood, B., and Iacopi, F., Graphitized silicon carbide microbeams: wafer-level, self-aligned graphene on silicon wafers, *Nanotechnology*, **25**, 325301 (2014).

106. Lee, C., Wei, X., Kysar, J. W., and Hone, J., Measurement of the elastic properties and intrinsic strength of monolayer graphene, *Science*, **321**, 385–388 (2008).

107. Martin-Olmos, C., Rasool, H. I., Weiller, B. H., and Gimzewski, J. K., Graphene MEMS: AFM probe performance improvement, *ACS Nano*, **7**, 4164–4170 (2013).

108. Ahmed, M., Khawaja, M., Notarianni, M., Wang, B., Goding, D., Gupta, B., Boeckl, J. J., Takshi, A., Motta, N., and Saddow, S. E., A thin film approach for SiC-derived graphene as an on-chip electrode for supercapacitors, *Nanotechnology*, **26**, 434005 (2015).

Chapter 2

Graphene Synthesized on Cubic-SiC(001) in Ultrahigh Vacuum: Atomic and Electronic Structure and Transport Properties

V. Yu. Aristov,[a,b,c] O. V. Molodtsova,[b,d] and A. N. Chaika[a]

[a]*Institute of Solid State Physics RAS, Chernogolovka, Moscow District 142432, Russia*
[b]*Deutsches Elektronen-Synchrotron DESY, 22607 Hamburg, Germany*
[c]*Institut für Experimentelle Physik, TU Bergakademie Freiberg, 09596 Freiberg, Germany*
[d]*National Research University of Information Technologies, Mechanics and Optics, 197101 Saint Petersburg, Russia*
aristov@issp.ac.ru; olga.molodtsova@desy.de; chaika@issp.ac.ru

2.1 Introduction

Graphene possesses astonishing electronic properties [1–6], which are highly promising for use in electronic and photonic devices. For example, perspectives of technological applications in photosensors, transparent electrical contacts, and memory cells have attracted

---

Growing Graphene on Semiconductors
Edited by Nunzio Motta, Francesca Iacopi, and Camilla Coletti
Copyright © 2017 Pan Stanford Publishing Pte. Ltd.
ISBN 978-981-4774-21-5 (Hardcover), 978-1-315-18615-3 (eBook)
www.panstanford.com

great attention during recent years [7–13]. Furthermore, its unique transport properties make graphene a very appealing candidate for replacing traditional silicon-based by novel nanoscaled carbon-based electronics and developing the beyond-complementary-metal-oxide-semiconductor (CMOS) technologies.

Strictly speaking, the word "graphene" refers to a single layer of sp^2-hybridized carbon atoms tightly packed in a honeycomb crystal lattice by three in-plane σ bonds per atom. Nevertheless, the term "graphene" is often employed to describe bilayer and few-layer graphene, which are also considered as peculiar kinds of 2D crystals [2]. This chapter is focused on the synthesis and characterization of single- and few-layer graphene on cubic-SiC(001)/Si(001) wafers in ultrahigh vacuum (UHV). After a brief overview of different methods of graphene fabrication, we review the published studies of the atomic and electronic structure of the graphene overlayers on SiC(001) surfaces. At the end of the chapter, the results demonstrating transport properties of graphene synthesized on the vicinal SiC(001) substrate are presented.

2.2 Synthesis of Few-Layer Graphene

2.2.1 Methods of Graphene Fabrication

During the recent years, numerous promising procedures of graphene fabrication have been reported. Historically, the first methods of graphene fabrication involved handmade processes, such as mechanical and chemical exfoliation [1, 14–17]. Although these methods are hardly suitable for technologies, they can provide the highest-quality samples for fundamental research. For example, the exceptional properties of a 2D electron gas in single-layer graphene have been experimentally observed for the first time during the studies of ultrathin graphite flakes mechanically exfoliated from highly oriented pyrolytic graphite (HOPG) crystals [5, 6]. Nevertheless, the small size of the graphene sheets and necessity of manipulation by micrometer-scaled flakes in isolation from one another prevent the utilization of this method for mass production of graphene compatible with electronic technologies. Other synthesis procedures are usually based on carbonization of metallic and semiconducting surfaces either in vacuum or in an atmosphere of different gases.

Graphitization of single-crystal metal surfaces at specific environment and substrate temperatures has been known at least since 1965. The formation of ultrathin graphite films was first detected during investigations of Pt and Ru single-crystal surfaces [18–22]. Graphitization of metal surfaces was also observed in manufacturing processes using heterogeneous catalysts [23, 24].

Self-assembly of carbon atoms into a honeycomb lattice can be anticipated on threefold-symmetric surfaces. That was confirmed for graphene grown by thermal chemical vapor deposition (CVD) on transition metal (Ni(111), Ir(111), etc.) [25–27] and semiconductor surfaces [28, 29], as well as for graphene synthesized using plasma-enhanced CVD [30–33] and graphite oxide reduction [34–37]. These methods also have disadvantages, preventing the mass production of graphene for micro- and nanoelectronic devices. For example, graphene layers fabricated by a catalytic reaction or the CVD method on metal surfaces cannot be used for electronic applications because of substrate conductivity. Although high-quality, large-size graphene overlayers can be grown by CVD on metal surfaces, the graphene sheets must be transported from conducting onto insulating substrates for device fabrication. This transfer step is usually accompanied by a contact of the few-layer film with an aggressive chemical environment, which can lead to a substantial modification of the unique properties of graphene. Therefore, a better option would be direct synthesis on single-crystal surfaces of wide-bandgap semiconductors (e.g., BN or SiC).

Among the most promising semiconducting substrates having the surface structure suitable for graphene synthesis we can mention hexagonal (α-SiC) 6H- and 4C-SiC(0001) and cubic (β-SiC) 3C-SiC/Si(111) surfaces [38–41]. The graphitization of hexagonal silicon carbide was first reported in 1975 by van Bommel et al. [42]. In 2001 the group of de Heer elaborated the procedure of creating planar graphene layers by heating SiC wafers to temperatures exceeding 1300°C. Inspired by the development of new 2D electronics, in 2004 they demonstrated the 2D electron gas properties of the charge carriers in ultrathin epitaxial graphite films on 6H-SiC(0001) subjected to an electric field [43]. The fabrication of few-layer graphene on α-SiC is based on relatively simple thermal decomposition reaction resulting in silicon atoms' sublimation and formation of carbon-enriched surface layers [38, 44]. Further annealing of the carbon-enriched α-SiC surface at high temperatures leads to its graphitization. This method was proposed for the synthesis of homogeneous, epitaxial, wafer-scale few-layer graphene for technological applications [45–47].

Noteworthy is that the electronic properties of the few-layer graphene synthesized using this method on α-SiC are similar to that of freestanding single-layer graphene. As an example, Fig. 2.1 displays the band structure of the 11-layer graphene film grown on the 6H-SiC and studied experimentally using the angle-resolved photoelectron spectroscopy (ARPES) [48]. The ARPES measurements revealed nearly ideal linear dispersions at the K-points, typical of freestanding graphene monolayer (Fig. 2.1a).

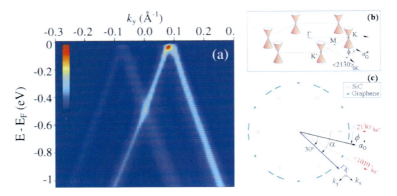

Figure 2.1 (a) Band structure of an 11-layer C-face graphene film grown on the 6H-SiC measured by ARPES at $\hbar\omega$ = 30 eV and T = 6 K. The scan in k_y is perpendicular to the SiC <10-10>$_{SiC}$ direction at the K-point (b, c). Two nearly ideal linear dispersions are visible. (b) 2D Brillouin zone of graphene near E_F showing the six Dirac cones. The graphene reciprocal lattice vector $\mathbf{a}^*_G$ (and therefore the cones) is rotated by φ = 30° relative to the SiC <21-30> direction. (c) Schematic diffraction pattern of graphene grown on SiC(000-1). The SiC and the graphene patterns are indicated by open circles and green ovals, respectively. Diffuse graphene arcs on the C face are centered at φ = 0°. Reprinted (figure) with permission from Ref. [48]. Copyright (2009) by the American Physical Society.

Since silicon carbide is a wide-bandgap semiconductor (from 2.4 eV for 3C-SiC to 3.3 eV for 4H-SiC) this method gives a possibility to grow few-layer graphene on an almost insulating substrate. Therefore, the synthesized graphene overlayer does not need any transfer from one surface to another before processing devices. The high-temperature synthesis on the α-SiC surfaces in vacuum or an argon atmosphere at high pressure is one of the best methods developed so far for graphene fabrication on insulating substrates. Nevertheless, this method does not meet the requirements of industrial mass production because of the limited size and high costs of α-SiC wafers, which are usually cut from bulk single-crystal ingots with consequent polishing.

2.2.2   Growth of Cubic-SiC Epilayers on Standard Si Wafers

For many years, SiC has been a prime candidate for high-temperature electronic applications because of its large bandgap, high carrier mobility, and enhanced stability of the crystal lattice. For example, huge efforts were aimed at developing SiC-based high-temperature electronic devices for advanced turbine engines, geothermal wells, and other applications. Replacement of silicon by silicon carbide in integrated circuits (IC) would allow the improvement of the characteristics of the electronic devices and reduce their degradation at elevated temperatures. However, the development of SiC-based electronic technologies had been complicated by the lack of a reproducible process for mass production of low-cost single-crystal silicon carbide substrates. To address this issue, heteroepitaxial growth of cubic silicon carbide polytype (3C-SiC or β-SiC) on silicon wafers was proposed in the 1980s, when Nishino et al. [49] synthesized a single-crystal 3C-SiC(001) epilayer on a single-crystal Si(001) wafer using CVD and showed the feasibility of the growth of high-quality, uniform single-crystal cubic-SiC(001) layers on centimeter-size silicon substrates. Afterward, numerous works have been done to improve the synthesis of 3C-SiC(111) and 3C-SiC(001) epilayers on Si(111) and Si(001) substrates [50–52], respectively, and investigate their atomic structure and electronic properties.

   One of the first studies of the electronic structure of high-quality β-SiC(001) epilayers grown on Si(001) wafers was reported in 1987 by Hoechst et al. [53], who investigated the valence-band structure of β-SiC(001)/Si(001) with ARPES using synchrotron radiation. However, detailed investigations of the atomic structure and electronic properties of β-SiC(001) epilayers grown on Si(001) had been performed in several research centers during the 1990s. The results of these studies were summarized, for example, in reviews [54, 55].

2.2.3 Synthesis of the Epitaxial Graphene Layers on (111)- and (011)-Oriented Cubic-SiC Films Grown on Si Wafers

In contrast to α-SiC single-crystal wafers, thin films of β-SiC had not been considered as suitable substrates for graphene synthesis for quite a long time. This could be related to the cubic lattice of β-SiC, which appeared to be incommensurate with the graphene honeycomb lattice. Synthesis of epitaxial graphene (EG) layers on cubic-SiC thin films deposited onto silicon wafers, defined as graphene-on-silicon (GOS) technology, has been studied during the recent years [56–62]. Since the self-assembly of carbon atoms into a honeycomb lattice can be readily anticipated on threefold-symmetric surfaces, a greater number of studies have been carried out on SiC(111) and SiC(011) surfaces, while only a limited number of studies have been related to the 3C-SiC(001)/Si(001) system.

Investigations of the structural and electronic characteristics have been reported for graphene fabricated on single-crystalline 3C-SiC(111)/Si(111), [57, 59, 60, 63–67] 3C-SiC(111)/Si(011), [68–70], and 3C-SiC(111)/Si(001) [71] substrates. A number of studies have also been performed on single-crystalline bulk cubic-SiC(111) samples [72]. According to the published results, the graphitization of the 3C-SiC(111)/Si(111) system is analogous to that of the Si-terminated α-SiC single-crystal surfaces. In particular, the photoelectron spectroscopy (PES) data prove the existence of a buffer layer between the graphene overlayer and the SiC(111) substrate, which is similar to graphene grown on α-SiC surfaces.

As an example, Fig. 2.2a demonstrates the C 1s core-level photoemission spectra taken from thermally graphitized 3C-SiC(111)/Si(111), 3C-SiC(001)/Si(001), and 3C-SiC(011)/ Si(011) substrates [57]. In all the spectra, the peak related to sp^2

carbon atoms, proving the formation of the graphene overlayer, can be detected along with the SiC bulk peak (located at lower binding energies [BEs]). In general, all three spectra show similar shape. Nevertheless, in the spectrum recorded from the graphene grown on the SiC(111)/Si(111) substrate one can see a more intense shoulder at higher BEs. This is indicated by peak I having approximately 1 eV higher BE relative to the graphene sp^2 component. This peak suggests the existence of the interfacial layer and the charge transfer within the incommensurate graphene/SiC(111) system. A similar charge transfer has been reported for EG on Si-face/6H-SiC(0001), which is accompanied by a Bernal stacking of the graphene layers [43, 72, 73]. Therefore, the presence of the reactive interfacial component in the C 1s core-level spectrum was explained by the formation of the Bernal-stacked few-layer graphene on the 3C-SiC(111)/Si(111) surface, while non-Bernal stacking was suggested for the graphene grown on the 3C-SiC(001)/Si(001) and 3C-SiC(011)/Si(011) surfaces. According to other studies [69], graphene synthesis on the SiC(111)/Si(011) wafers was also not accompanied by the formation of the reactive buffer layer. Similar behavior of the C 1s core-level spectra was observed for graphene synthesized on 3C-SiC(111) and 3C-SiC(001) epilayers deposited onto (111) and (001) microsteps, respectively, fabricated on the single-crystalline Si(001) surface [73].

The interface structure between graphene and bulk silicon carbide significantly affects the electronic properties of the EG on SiC crystals [43, 48, 74]. Indeed, depending on the surface termination, the EG grown on Si-terminated and C-terminated surfaces of the α-SiC bulk crystals demonstrates semiconducting and metallic properties, respectively [43, 73–75]. As shown by Suemitsu and Fukidome [57], a similar dependence on the Si substrate orientation can also be expected in GOS.

The principal difference between the graphene overlayers on the 3C-SiC(111)/Si(111) and other surface orientations was also confirmed by the Raman data. As an example, Fig. 2.2b shows that G' peak in the spectra measured from the EG on 3C-SiC(111)/Si(111) splits into multiple components (reactive and nonreactive), which is not the case for the other two cubic-SiC/Si orientations [57].

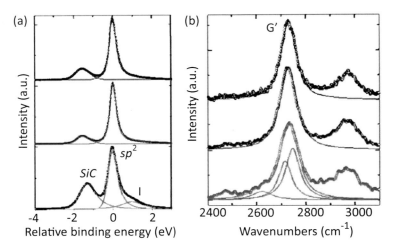

Figure 2.2 (a) Comparison of the C 1s core-level spectra taken from EG on 3C-SiC/Si(111) (bottom), 3C-SiC/Si(001) (middle), and 3C-SiC/Si(011) (top). (b) Raman spectra of EG on 3C-SiC/Si(111) (bottom), 3C-SiC/Si(001) (middle), and 3C-SiC/Si(011) (top) substrates. Reproduced from Ref. [57]. Copyright © IOP Publishing. Reproduced with permission. All rights reserved.

2.3 Synthesis and Characterization of Continuous Few-Layer Graphene on Cubic-SiC(001)

Among the SiC/Si wafers, 3C-SiC(001) is the most appealing substrate for developing new, graphene-based technologies. 3C-SiC(001) films can be readily grown as thin single-crystal epilayers on standard large-size (up to 300 mm in diameter) Si(001) wafers, generally used for IC fabrication, and now are commercially available [76]. The feasibility of graphene synthesis on low-cost, technologically relevant 3C-SiC(001) wafers in UHV [56, 77–92] and in an argon atmosphere [62] has been demonstrated recently. The synthesis on 3C-SiC(001) surfaces represents a realistic method for the mass production of graphene layers suitable for electronic applications and compatible with existing silicon technologies. Therefore, graphene overlayers grown on such insulating substrates can be considered as appropriate materials for future graphene-based electronics. The following sections summarize the published studies related to the synthesis and characterization of graphene on low-index and vicinal SiC(001) surfaces using various surface science techniques.

2.3.1 Step-by-Step Characterization of SiC(001) Surface during Graphene Synthesis in Ultrahigh Vacuum

Uniform single- and few-layer graphene coverages can be fabricated on SiC(001)/Si(001) wafers in UHV using Si-atom sublimation followed by high-temperature surface graphitization [56, 77–92]. The first steps toward graphene synthesis on SiC(001)/Si(001) wafers are related to the removal of the protective silicon oxide layer and the fabrication of a contaminant-free SiC(001)1×1 structure. These first stages of graphene synthesis help to avoid surface contamination, which can modify the atomic structure and properties of the graphene grown on SiC(001). SiC(001)1×1 reconstruction is generally fabricated after outgassing the sample holder and flash-heating the SiC(001)/Si(001) wafers for several seconds at 1000°C–1100°C. Then, fabrication of a graphene overlayer includes the deposition of several monolayers of silicon atoms onto the clean, carbon-rich SiC(001)1×1 surface and annealing at gradually increasing substrate temperatures in the range of 700°C–1350°C. Figure 2.3 summarizes the results of the core-level PES, low-energy electron diffraction (LEED), and scanning tunneling microscopy (STM) studies of the SiC(001) surface atomic structure as a function of annealing temperature.

Figure 2.3a–g shows the results of the PES experiments with the real-time control of the C 1s core-level spectrum shape in the course of the direct current sample heating [93]. During the in situ PES measurements a current was applied to heat the sample up to 1350°C (Fig. 2.3a). The C 1s core-level spectra were taken in a snapshot mode during the SiC(001)/Si(001) sample heating and graphene synthesis. More than 900 spectra were recorded with an acquisition time of 1 s/spectrum during 15 min, using a photon energy of 750 eV. Six core-level spectra taken at different temperatures during the graphene synthesis are shown in Fig. 2.3b–g. Despite the short time of spectrum acquisition and high sample temperature, two C 1s peak components can be distinguished in the spectra, which change their relative intensity at different preparation stages. Note that absolute BEs of individual components can slightly change from true values because of the voltage applied across the SiC/Si wafer. Therefore, only the relative BEs of the two different C 1s components on Fig. 2.3 should be analyzed.

In the first stage of the sample heating (Fig. 2.3b) one can see a strong peak corresponding to the carbon atoms in the bulk of the SiC crystal. Increasing the temperature (Fig. 2.3b–e) gives rise to a small additional component shifted by 1.55 eV to a higher BE, which starts to grow at temperatures above 1250°C. At the same time the relative intensity of the bulk component decreases with the temperature going above 1250°C. The change of the C 1s core-level shape corresponds to the sublimation of Si atoms and graphitization of the top surface layers at high temperatures. In the last stage (Fig. 2.3g), the temperature of the sample reached 1350°C, which is close to the silicon melting point. At this temperature the carbon–carbon bonds undergo a reconstruction to sp^2 hybridization, corresponding to graphene honeycomb lattice formation. The ex situ LEED measurements proved the existence of a graphene overlayer on the SiC(001)/Si(001) wafer used for the PES experiments with real-time control of the surface composition.

Figure 2.3h–l shows the results of the step-by-step LEED and STM characterization of the SiC(001) surface atomic structure after direct current sample heating in UHV [84, 85]. The LEED and STM experiments prove that changes of the relative concentrations of silicon and carbon atoms during sample heating lead to consecutive fabrication of various reconstructions in accordance with Refs. [94–100]. The LEED and STM data stress several crucial stages of graphene synthesis on 3C-SiC(001). In the first stage, long-term annealing at temperatures in the range of 700°C–1000°C is important to fabricate a uniform, Si-rich SiC(001)3×2 reconstructed surface (Fig. 2.3h) with large (001)-oriented terraces. Increasing the annealing temperature from 1000°C to 1250°C causes the surface to undergo consecutive silicon-terminated 5×2 (Fig. 2.3i), c(4×2) (Fig. 2.3j), 2×1, and carbon-terminated c(2×2) (Fig. 2.3k) reconstructions. The LEED and STM data shown in Fig. 2.3h–l were obtained after consecutively heating the same SiC(001)/Si(001) sample to 1000°C, 1150°C, 1200°C, 1250°C, and 1350°C and cooling it down to room temperature. However, the precise temperatures of the surface phase transitions could be diverse at slightly different pressures in the UHV chambers used for graphene synthesis. Note that UHV STM studies of the c(4×2) reconstruction usually revealed missing Si atoms in the surface layer, while the c(2×2) structure was typically decorated by excessive carbon atoms and atomic chains elongated in one of the orthogonal <110> directions [85]. The absence of impurities on the SiC(001) surface during graphene synthesis was confirmed by the fabrication of the c(4×2) reconstruction (Fig. 2.3j), which is very sensitive to surface contamination. The c(4×2) structure usually transforms into the 2×1 phase after several hours because of exposure to background hydrogen, even in UHV [98]. The extra carbon atoms on the SiC(001)-c(2×2) reconstruction are important to transform it into the more densely packed honeycomb lattice (Fig. 2.3l). According to LEED and STM studies, the most uniform graphene overlayers on SiC(001) can be obtained after flash heating (10–20 s) of the c(2×2) reconstruction at 1350°C with postannealing at 600°C–700°C. The short flash method is similar to that used for the growth of graphene on α-SiC [74, 101, 102]. The LEED pattern shown in Fig. 2.3l clearly reveals sharp (1×1) SiC substrate spots and a ring of 12 double-split graphene spots related to the formation of the network of graphene nanodomains with several preferential lattice orientations [84, 85]. Since the probing area in the conventional LEED experiments is of the millimeter size, the sharpness of the diffraction spots proves a few-layer thickness of graphene on SiC(001) and the presence of all preferential lattice orientations on large sample areas.

The described procedure was utilized for the fabrication of few-layer graphene on 3C-SiC(001) in various works [56, 81–92]. The exact number of graphene layers on SiC(001) reported by different authors, usually, did not exceed three atomic layers. It can be assumed that the growth of few-layer graphene on SiC(001) at pressures in the low 10^{-10} mbar region is a self-limiting process. Most probably, multilayer graphene coverages cannot be fabricated on well-outgassed SiC(001)/Si(001) wafers at pressures below 2×10^{-10} mbar. The SiC(001) graphitization can be extremely dependent on the sample cleanliness, pressure in the vacuum chamber, etc. For these reasons, it can be difficult to obtain a uniform monolayer graphene coverage on millimeter-scale SiC(001) samples although the possibility to synthesize a uniform few-layer coverage has been proved by a variety of experimental techniques [84].

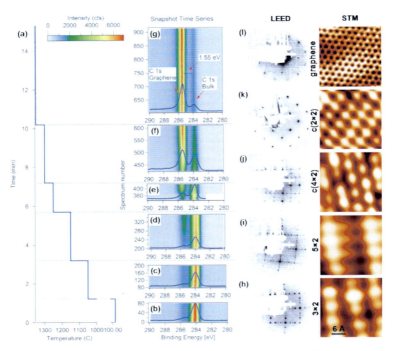

Figure 2.3 (a–g) In situ core-level PES studies of the SiC(001)/Si(001) atomic structure during sample heating in UHV. (a) Temperature of the sample during PES measurements. (b–g) Time evolution of C 1s core-level spectra recorded in the snapshot regime during sample heating. A single snapshot spectrum taken in the corresponding temperature interval shown in (a) is presented. (h–l) Evolution of the SiC(001) surface atomic structure with increasing temperature registered by LEED and STM. The 3×2, 5×2, c(4×2), and c(2×2) reconstructions are consecutively formed on the SiC(001) surface in the temperature range of 800°C–1300°C before the synthesis of few-layer graphene. Panels (a–g) adapted from Ref. [93], Copyright (2015), with permission from Elsevier.

Figure 2.4 shows the results of the near-edge X-ray absorption fine structure (NEXAFS) and PES studies of graphene on SiC(001) conducted at the photon energies in the range of 280–400 eV [56], corresponding to enhanced surface sensitivity. Figure 2.4a shows the C 1s photoemission spectra taken at $h\nu$ = 400 eV from the SiC(001)3×2 and SiC(001)-c(2×2) reconstructions and graphene/SiC(001). For the Si-rich SiC(001)3×2 surface, all carbon atoms occupy equivalent bulk sites. Hence, only one (bulk) component with a BE of 282.9 eV is present, which is analogous to the lower BE component observed in the real-time PES experiments presented in Fig. 2.3b. Surface carbon atoms give rise to a second PE component shifted by 1.55 eV toward a higher BE in the core-level spectra taken from the C-rich SiC(001)-c(2×2) and graphene/SiC(001) surfaces. The spectra of the SiC(001) surface reconstructions obtained in the course of graphene synthesis (Figs. 2.3 and 2.4) are in excellent agreement with the earlier studies [103].

To verify the origin of both components in the core-level spectra, the photon energy was varied between 315 and 350 eV (Fig. 2.4b), where the mean free path of the electrons changes, reaching its minimum at about hv = 325 eV. The bulk component is suppressed at this photon energy, whereas the surface peak remains unaffected. The energy position and the full width at half-maximum (FWHM) of the surface component are typical of graphite/graphene. The C 1s spectra taken from the graphene/SiC(001) system demonstrate the absence of the reactive components, corresponding to the formation of a buffer layer that is in agreement with the other PES studies of graphene/SiC(001) and graphene/SiC(011) systems [57]. The absence of the reactive component in the C 1s spectra (Fig. 2.4) proves the quasi-freestanding character of graphene synthesized on SiC(001)/Si(001) wafers. This is in contrast with the results obtained on 3C-SiC(111) and hexagonal SiC, where the graphene overlayer starts to grow after the formation of the reactive phase.

The NEXAFS spectra taken with the linearly polarized photons from graphene/SiC(001) samples at three different photon incidence angles (Fig. 2.4c) are characterized by sharp resonances at 285.3 and 291.6 eV and a broad structure at 292.7 eV. These features are typical of pristine graphite, which are related to the π^*, σ_1^*, and σ_2^* resonances, respectively [104]. For the 2D π-conjugated systems C 1s $\rightarrow \pi^*$ or 1s $\rightarrow \sigma^*$ transition probabilities are maximal for the electrical vector perpendicular or parallel to the surface, respectively, which is in agreement with the observed polarization dependence (Fig. 2.4c). At grazing incidence, when the direction of the electrical vector is close to the surface normal, the intensity of the π^* resonance is strongly enhanced. It decreases with an increasing incident angle (Θ = 45°) and is almost suppressed for normal incidence. The σ^* resonances reveal the opposite angular dependence, showing the maximal relative intensity at normal incidence. The observed polarization dependence is typical of single- and few-layer graphene and graphite. Noteworthy is that the observed C 1s NEXAFS spectral line shapes are not substantially affected by the interaction with the underlying substrate [105], independently proving a quasi-freestanding character of the graphene overlayer, not substantially interacting with the substrate.

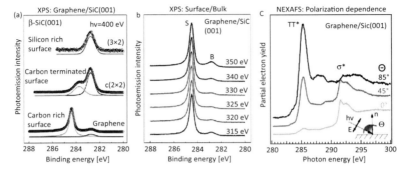

Figure 2.4 Core-level photoemission and NEXAFS studies of graphene/SiC(001). (a) C 1s photoemission taken from SiC(001) at different stages of the sample preparation using a photon energy of 400 eV. The deconvolution of the C 1s spectrum taken after graphene formation shows only two components corresponding to the overlayer and SiC crystal. (b) C 1s spectrum taken from graphene/SiC(001) as a function of photon energy. As $h\nu$ is tuned away from the most surface-sensitive regime at 325 eV, component B rises, which proves its bulk origin. The binding energy of the surface component S corresponds to graphene. (c) C 1s NEXAFS spectra recorded at three different incidence angles Θ of the linearly polarized light. The spectra taken at grazing and normal incidence suggest that the π* orbitals are oriented parallel to the surface normal and the σ* orbitals are parallel to the surface, as expected for graphene/graphite.

Additional proof of the SiC(001) surface graphitization during high-temperature annealing in UHV was obtained in core-level PES and Raman spectroscopy experiments [77]. Figure 2.5 shows the Raman spectra measured in different stages of a few-layer graphene growth on SiC(001). The measurements were performed for two different stages of the surface graphitization after the preparation of the SiC(100)-c(2×2) phase (namely 1.5 and 2.8 atomic layers over the SiC substrate). The Raman spectrum taken from the c(2×2) reconstruction reveals several peaks in the 1000–2000 cm^{-1} range. The peak at 1516 cm^{-1} is considered to be an overtone of the L-point optical phonon. The Raman signals from the two different graphene overlayers show prominent peaks slightly above 1600 cm^{-1} (G) and 2700 cm^{-1} (2D), which give evidence of carbon sp^2 reorganization.

The blue-shifted position of the 2D peak at 2732 cm^{-1} can indicate a compressive strain of the graphene layer synthesized at high temperatures and cooled down to room temperature. The observed Lorentzian shape of the 2D feature is typical of EG layers. However, the FWHM of the 2D peak is around 80 cm^{-1}, exceeding the values known for ideal freestanding graphene. The broadening can be attributed to the presence of a large number of defects in the few-layer graphene synthesized on 3C-SiC(001). According to the ratio of the intensities of the G and D peaks in the Raman spectra, the authors estimated that the average distance between defects in graphene/SiC(001) should be of the order of 10 nm [77]. Recent STM, low-energy electron microscopy (LEEM) and ARPES studies of the uniform trilayer graphene grown on SiC(001) [84, 85] disclosed the structure of the graphene overlayer on SiC(001) at the atomic level.

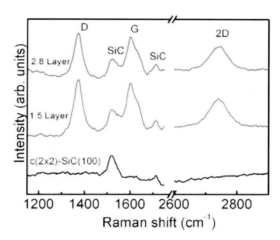

Figure 2.5 Micro-Raman characterization of graphene/SiC(100): comparison of Raman spectra taken on the c(2×2) surface reconstruction and on two graphene layers. Reprinted (figure) with permission from Ref. [77]. Copyright (2011) by the American Physical Society.

2.3.2 Atomic and Electronic Structure of the Trilayer Graphene Synthesized on SiC(001)

Figure 2.6 shows the large-area STM images taken from the SiC(001) surface before and after the trilayer graphene synthesis [84, 85].

The extra carbon atoms on the SiC(001)-c(2×2), necessary for the honeycomb lattice formation are resolved as bright protrusions on terraces (Fig. 2.6a). The single atomic steps are clearly seen in Fig. 2.6a. STM studies [85] revealed that the root mean square (RMS) roughness of micrometer-scale (1×1 μm^2) SiC(001)-c(2×2) STM images exceeded 1.5 Å only in rare cases. The roughness analysis for a smaller (100×100 nm^2) surface area inside one of the (001)-oriented terraces is shown in Fig. 2.6d. The RMS for this surface region is below 1.0 Å. The heights of the single atomic step on the STM images taken from the SiC(001)-c(2×2) and graphene/SiC(001) surfaces were in good agreement with the well-known cubic-SiC lattice parameters. This is illustrated by the cross-section of the STM image measured near the monatomic step on graphene/SiC(001) (Fig. 2.6k). Therefore, the RMS values shown in Fig. 2.6 for the SiC(001) surface before and after graphene synthesis, presumably, correspond to the actual roughness of the top surface layer.

Thorough STM studies [85] proved the absence of bare silicon carbide regions on graphene/SiC(001) samples. For example, STM images shown in Figs. 2.6b and 2.6c were taken at the bias voltages of −1.0 and −0.8 V, respectively, that is in the bandgap of 3C-SiC (2.4 eV). STM experiments at such low biases would not be possible on SiC(001) surface reconstructions (Fig. 2.3). Figures 2.6b and 2.6c demonstrate that STM imaging is stable even on micrometer-scale graphene/SiC(001) surface areas containing defects, namely antiphase domain (APD) boundaries (Fig. 2.6c) and multiatomic steps (Fig. 2.6b). Although the resolution is limited near the defects and actual topography cannot be seen in the defect regions because of the tip effects, the absence of a jump-to-contact at such small bias voltages confirms the continuity of the graphene overlayer on cubic-SiC(001), which is not broken by multiatomic step edges and APD boundaries.

As STM images in Figs. 2.6g and 2.6h illustrate, the top graphene layer consists of nanodomains connected to one another through the domain boundaries. According to the STM studies [84, 85], the nanodomain boundaries (NBs) are preferentially aligned with the two orthogonal <110> directions of the SiC crystal lattice, as indicated in Figs. 2.6g and 2.6h. The domains are elongated in the [110] and [1–10] directions on the right and left side of the APD boundary, respectively (Fig. 2.6c). The nanodomains on the SiC(001) substrate have lengths varying between 20 and 200 nm and widths

Synthesis and Characterization of Continuous Few-Layer Graphene on Cubic-SiC(001) | 43

in the range of 5–30 nm, although wider nanodomains were also observed.

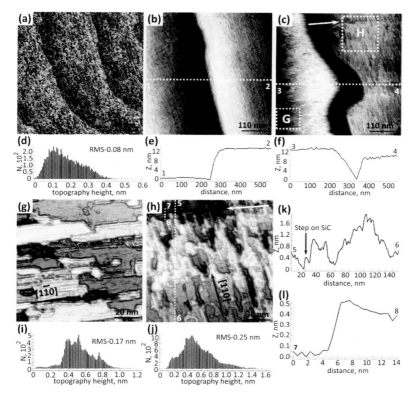

Figure 2.6 Large-area STM images of SiC(001)-c(2×2) (a) and graphene/SiC(001) (b, c, g, h). The images in panels (b) and (c) illustrate the continuity of the graphene overlayer near the multiatomic step (b) and the APD boundary (c). The images in panels (g) and (h) emphasize the nanodomains elongated along the [1–10] (g) and [110] directions (h) observed on the left and the right side, respectively, of the APD boundary in (c). The images in panels (g) and (h) were measured from the surface areas labeled G and H in panel (c). The STM images were measured at $U = -3.0$ V and $I = 60$ pA (a), $U = -1.0$ V and $I = 60$ pA (b), $U = -0.8$ V and $I = 50$ pA (c), $U = -0.8$ V and $I = 60$ pA (g), and $U = -0.7$ V and $I = 70$ pA (h). The white arrows in panels (c) and (h) indicate a single atomic step on the SiC substrate. (d, i, j) Roughness analysis of the STM images in panels (a), (g), and (h). The histograms were calculated from surface areas of the same size (100 × 100 nm^2) for direct comparison of the surface roughness before and after graphene synthesis. (e, f, k, l) Cross sections (1–2), (3–4), (5–6), and (7–8) taken from the images in (b), (c), and (h). Reproduced from Ref. [85]. Copyright © IOP Publishing. Reproduced with permission. All rights reserved.

Graphene Synthesized on Cubic-SiC(001) in Ultrahigh Vacuum

STM studies revealed that individual graphene domains on SiC(001) possess a rippled morphology, which leads to a RMS roughness of micrometer-scale STM images having the order of several angstroms (e.g., see Figs. 2.6i and 2.6j, which demonstrate an RMS of 1.7 Å and 2.5 Å for 100×100 nm^2 surface areas, respectively). Note that the roughness calculated from images of the same size is approximately a factor of 2 larger for graphene/SiC(001). The roughness of 1×1 µm^2 STM images of graphene/SiC(001) was usually between 2.5 and 4.5 Å, which is typical of freestanding graphene [106, 107]. However, despite the surface roughness, single atomic steps under the trilayer could be easily resolved with STM, as Figs. 2.6c and 2.6k illustrate. This fact, as well as the high intensity and sharpness of the SiC substrate spots in the LEED patterns (e.g., Fig. 2.3l) suggest that the thickness of the graphene overlayer for the studied sample does not exceed several monolayers. The number of graphene layers was precisely identified using the LEEM [84].

Figure 2.7a shows a typical LEEM micrograph ($E = 3.4$ eV) of the graphene/SiC(001) surface demonstrating APD boundaries and uniform contrast throughout the probed 20 µm surface area. The uniform contrast in Fig. 2.7a proves that despite all the defects of the cubic-SiC substrate (Fig. 2.6), the thickness of the synthesized graphene is uniform across the probed surface area [41]. The number of graphene layers can be deduced from the reflectivity curves acquired in a 7 eV energy window [108, 109]. As Fig. 2.7h illustrates, there are three distinct minima in the reflectivity spectra taken from different surface areas, which correspond to the uniform triple layer. The graphene coverage is very homogeneous all over the surface, and $I–V$ curves are almost identical in different surface regions, including the areas with different contrast in dark-field (DF) LEEM images (e.g., areas 1–3 in Fig. 2.7b).

The micro-LEED pattern taken from a 5 µm area reveals 12 double-split diffraction spots from the graphene trilayer and well-resolved spots from the SiC(001) substrate (Fig. 2.7d). DF LEEM images of the same surface area (Figs. 2.7b and 2.7c) and micro-LEED patterns taken from different 1.5 µm areas, shown in Figs. 2.7e and 2.7f, demonstrate that the 12 double-split spots originate from different micrometer-scale surface areas, producing 90°-rotated micro-LEED patterns with 12 nonequidistant spots (Figs. 2.7e and 2.7f). These areas appear as white and dark in corresponding reflexes (Figs. 2.7b and 2.7c).

Synthesis and Characterization of Continuous Few-Layer Graphene on Cubic-SiC(001) | 45

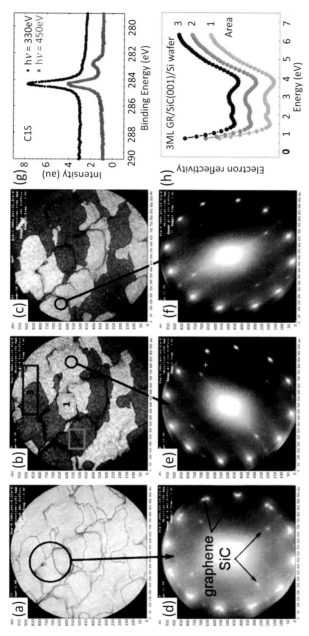

Figure 2.7 (a) 20 μm LEEM micrograph, recorded with an electron energy of 3.4 eV, proving the uniform thickness of the graphene synthesized on SiC(001)/Si(001) wafers. (b,c) DF LEEM images from different diffraction spots, shown in panels (e) and (f), demonstrating the contrast reversal on micrometer-scale areas with two rotated graphene domain families. (d–f) Micro-LEED patterns from the surface areas shown in panels (a–c). The sampling areas are 5 μm (a) and 1.5 μm (b, c); $E = 52$ eV. (g) Micro-PES C 1s core-level spectra taken in the normal incidence–normal emission geometry at different photon energies. The sampling area is 10 μm. (h) Electron reflectivity spectra recorded for the different surface regions 1, 2, and 3, as labeled in panel (b), where the number of dips in the spectra identifies regions 1–3 as trilayer graphene. Reproduced from Ref. [84] with permission from Tsinghua and Springer.

The C 1s spectra shown in Fig. 2.7g reveal only two narrow components with BEs corresponding to the SiC substrate (lower BE) and graphene trilayer (higher BE), which proves the absence of a buffer layer and the weak interaction of the synthesized overlayer with the substrate. ARPES and STM studies [84, 85, 92] also proved the quasi-freestanding character of the trilayer graphene on SiC(001) and disclosed the origin of the 12 double-split diffraction spots in the LEED patterns (Figs. 2.3 and 2.7). Note that micro-LEED patterns of graphene/SiC(001) demonstrating split diffraction spots were first reported by Ouerghi et al. [77] and ascribed to a systematic misorientation of two layers in a bilayer graphene. However, later atomically resolved STM studies [84, 85] indicated that the observed 24 diffraction spots are related to nanodomains with four preferential lattice orientations in the nanostructured few-layer graphene.

Atomically resolved STM images in Figs. 2.8a and 2.8b show nanodomains elongated in the [110] and [1–10] directions, respectively, and clarify the atomic structure of the graphene domain network on SiC(001). The 2D fast Fourier transform (FFT) patterns of the atomically resolved STM images (the inset in Fig. 2.8b) consist of two systems of spots (indicated by hexagons), which are related to two graphene lattices rotated by 27° relative to one another. It can be noted that this misorientation angle between neighboring domains' lattices was also observed in the electron microscopy studies of a suspended polycrystalline graphene [110]. The graphene/SiC(001) domains' lattices are preferentially rotated by ±13.5° from the [110] and [1–10] directions, while NBs are very close to these directions. These two families of 27°-rotated domains are themselves rotated by 90° relative to one another and produce two systems of 12 nonequidistant spots in the FFT (Fig. 2.8b) and micro-LEED patterns (Figs. 2.7e and 2.7f). The sum of two 90°-rotated patterns with 12 nonequidistant spots produce the LEED pattern of graphene/SiC(001) with 12 double-split spots, as the models shown in Figs. 2.8c, 2.8d, and 2.8e illustrate. These two orthogonal 27°-rotated domain families are resolved as horizontal and vertical nanoribbons in STM images, as shown in Figs. 2.6g and 2.6h.

Synthesis and Characterization of Continuous Few-Layer Graphene on Cubic-SiC(001) | 47

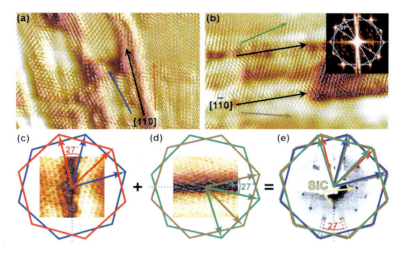

Figure 2.8 (a, b) 19.5 × 13 nm² atomically resolved STM images of the nanodomains elongated along the [110] (a) and [1–10] directions (b). The images were taken in different surface areas at $U = -10$ mV and $I = 60$ pA. The inset in (b) shows an FFT pattern with two 27°-rotated systems of spots. (c–e) Models explaining the origin of the 24 diffraction spots in the LEED patterns of graphene/SiC(001) in Figs. 2.3 and 2.7. Insets in panels (c) and (d) are STM images of the <110>-directed domain boundaries. The four differently colored hexagons, red, blue, green, and brown, represent the four domain orientations, indicated by similarly colored arrows in (a) and (b). The inset in panel (e) shows a LEED pattern taken at $E_p = 65$ eV, demonstrating 1 × 1 substrate spots (highlighted by yellow arrows) along with 12 double-split graphene spots, indicated by one dotted arrow for each orientation. Reproduced from Ref. [85]. Copyright © IOP Publishing. Reproduced with permission. All rights reserved.

DF LEEM images in Figs. 2.7b and 2.7c distinguish the surface areas containing the two orthogonal nanodomain families. As a result, the LEED patterns, recorded from ~1.5 μm surface areas shown in Figs. 2.7e and 2.7f, contain 12 nonequally spaced spots, similar to the FFT pattern of the high-resolution STM image shown in Fig. 2.8b. If a large surface area is probed by LEED one can see the superposition of the two diffraction patterns rotated by 90° (Fig. 2.7d) in accordance with the model shown in Fig. 2.8c–e. The DF LEEM images taken from different reflexes in either of double-split spot show a reversed contrast and confirm that the 27°-rotated domain families typically cover micrometer-size surface regions (Figs. 2.7b and 2.7c). The formation of two rotated nanodomain families on SiC(001) can be related to intrinsic defects of the cubic-SiC crystal lattice (APD boundaries). However, these defects do not break the continuity and uniform thickness of the graphene coverage, as STM (Fig. 2.6) and LEEM data (Fig. 2.7) prove.

Atomic resolution STM studies of the trilayer graphene synthesized on SiC(001) revealed the rippled morphology of the domains (bright regions in Figs. 2.8a and 2.8b) and additional electron density modulations near the domain boundaries. It is known that electronic states at graphene boundaries can induce additional modulations in STM images [111] and increase the periodicity of the atomically resolved patterns on nanosized graphene domains [112]. A $(\sqrt{3} \times \sqrt{3})R30°$ modulation is discernible near the domain edges in Figs. 2.8a and 2.8b. The $(\sqrt{3} \times \sqrt{3})R30°$ hexagonal spots are also seen in the FFT pattern (the inset in Fig. 2.8b). STM images taken from the surface regions far from the domain boundaries reveal hexagonal (Fig. 2.9a) or honeycomb (Fig. 2.9d) lattices distorted by atomic-scale rippling typical of freestanding graphene. The cross section of the STM image shown in Fig. 2.9c reveals both random vertical corrugations because of the graphene overlayer roughness and regular oscillations with a period of ~2.5 Å and an amplitude of 0.1–0.2 Å corresponding to the graphene lattice. The cross section shown in Fig. 2.9b illustrates that the dimensions of the ripples are of the order of several nanometers laterally and 1 Å vertically, coinciding with values calculated for freestanding monolayer graphene [107]. Such values were also obtained in STM experiments on an exfoliated graphene supported by a SiO_2/Si substrate [113], which is known to be the highest-quality graphene for fundamental research [1, 6]. The apparent heights of the ripples in the STM experiments could be enhanced because of tip–sample interactions and the high flexibility of the graphene layer [113, 114]. The distortions of the carbon–carbon bond lengths in the top graphene layer on SiC(001) are illustrated in Fig. 2.9d–f. The image in Fig. 2.9d shows both the honeycomb lattice and atomic-scale rippling. The contrast in the STM images taken on the top of one of the ripples (Figs. 2.9e and 2.9f) was adjusted to enhance the bond length distribution in small domain regions, which can be considered as planar. The image in Fig. 2.9e demonstrates the random picometer-scale distortions of the honeycomb lattice. For clarity, a distorted hexagon is overlaid on the STM image in Fig. 2.9f. Although the apparent lengths of the carbon–carbon bonds in scanning probe microscopy experiments can be modified by tip–surface interaction [115], the values shown in Fig. 2.9f agree with the theoretical calculation for the freestanding graphene monolayer [107].

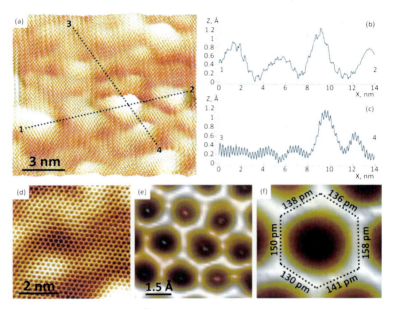

Figure 2.9 (a) 15 × 15 nm² STM image of trilayer graphene synthesized on SiC(001), illustrating atomic-scale rippling typical of freestanding graphene. The image was measured at U = 0.1 V and I = 60 pA. (b, c) Cross sections (1–2) and (3–4) from the image in panel (a) demonstrating the widths and heights of the ripples (b) and atomic corrugations with a periodicity of 2.46 Å (c). (d–f) STM images of the trilayer graphene demonstrating random picometer-scale distortions of the honeycomb lattice in the top layer. The images were measured at U = 22 mV and I = 70 pA (d) and U = 22 mV and I = 65 pA (e, f). One of the distorted hexagons is shown in (f) for clarity.

The flexibility and weak interaction of the graphene trilayer with the SiC(001) substrate is illustrated by Fig. 2.10(a–c), where a substantial change in the surface topography is detected with only a minor increase in the tunneling current, similar to effects observed on graphene/SiO$_2$ [116]. The roughness of the surface layer is modified by the tip–surface interaction after an increase in the tunneling current of only 33% (which corresponds to a change in the tip–sample distance of just several picometers). The images in Fig. 2.10 also show that some surface regions of nanostructured graphene (indicated by an arrow), which look like boundaries within the domain network at some tunneling parameters, can in fact be related to a bent single layer. The gap resistance–dependent STM experiments also showed that images with either hexagonal (Fig. 2.10d) or honeycomb (Fig. 2.10f) patterns can be resolved on the same surface areas at different tunneling parameters. This is illustrated in Fig. 2.10d–f by presenting the images measured at the same bias voltage and different tunneling currents. The contrast inversion is related only to the change in tunneling parameters used for STM imaging, not to a change in the overlayer thickness. The effect can be related to multiple-scattering effects in the tunneling gap [117], atomic relaxations, or modification of the orbital structure of the interacting tip and surface atoms at small distances [118]. This can lead to a situation where the tunneling current is mostly governed by the tip atoms in the second layer rather than by the apex atom closest to the surface. Therefore, at some specific distances a maximum in the tunneling current can be observed when the tip is located above the hollow sites and not above the true atomic positions in the honeycomb lattice.

The quasi-freestanding character of the trilayer graphene on SiC(001) is further confirmed by angle-resolved photoemission measurements of the π-band shown in Figs. 2.11 and 2.12. Since ARPES is a nonlocal method that sums up photoelectrons from a millimeter-scale sample area, the effective surface Brillouin zone of graphene on SiC(001) comprises Brillouin zones of all rotated domains, according to the sketch shown in Fig. 2.11a. This Brillouin zone can be directly derived from the LEED pattern displayed in Fig. 2.7d. The dispersions shown in Figs. 2.11b and 2.11c demonstrate that the Dirac cones sampled from all rotated domain variants are identical and their Dirac points are very close to the Fermi level. The similarity of the electronic structure of the rotated domains, the charge neutrality, and the absence of hybridization effects with SiC additionally emphasize the quasi-freestanding character of the synthesized trilayer graphene. The observed structure is consistent with the domains rotated by ±13.5° from the [110] and [1–10] directions and independently confirms that all 24 diffraction spots observed in the LEED experiments originate from the rotated domain network in trilayer graphene rather than from rotated layers in a bilayer structure [77].

Synthesis and Characterization of Continuous Few-Layer Graphene on Cubic-SiC(001) | **51**

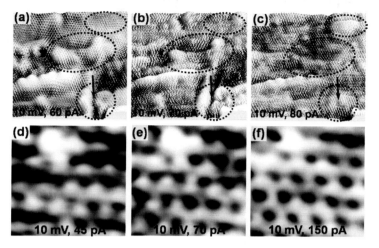

Figure 2.10 (a–c) Consecutive atomically resolved 18 × 18 nm² STM images of the same surface region of the trilayer graphene/SiC(001) measured at the same bias voltage and three different tunneling currents. The ovals indicate three areas where substantial change in the surface layer topography occurs because of tip–sample interaction. The arrows highlight a surface region appearing either as a domain boundary (a, b) or a bent graphene layer (c) at different tip–surface distances. (d–f) Gap resistance of the STM images of trilayer graphene on SiC(001), demonstrating the contrast reversal (from hexagonal to honeycomb pattern) with increasing tunneling current (decreasing tip–sample distance). The same defect is seen in the top-left corners of the 12 × 12 Å² images. The tunneling parameters are indicated on each particular frame. Reproduced from Ref. [85]. © IOP Publishing. Reproduced with permission. All rights reserved.

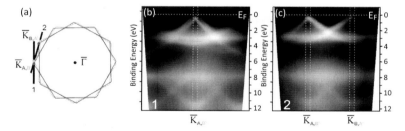

Figure 2.11 ARPES characterization of graphene grown on SiC(001). (a) Effective surface Brillouin zone as seen in ARPES due to the superposition of signals from the four rotated domain variants. The four domains are marked by letters A, B, A′, and B′. (b, c) Dispersion of π-band in graphene measured along directions 1 and 2 in (a). Reproduced from Ref. [84] with permission from Tsinghua and Springer.

The band structure of the graphene/SiC(001) sample measured along the $\overline{\Gamma}-\overline{K}$ direction of the surface Brillouin zone (long black line in Fig. 2.12a) reveals the characteristic dispersion of the π-band reaching the Fermi level. Figure 2.12b displays a dispersion of the π-band that back-folds at ∼2.5 eV BE and originates from the M point of the rotated graphene domain. To determine the energy of the Dirac point and to find out the charge doping of the trilayer graphene, the dispersions were measured in a detection geometry perpendicular to $\overline{\Gamma}-\overline{K}$ direction (short black line in Fig. 2.12a) where the interference of photoelectrons on graphene sublattices is suppressed and both sides of the Dirac cone are observed in photoemission [119]. The resulting ARPES data are shown in Fig. 2.12c. They reveal sharp linear dispersions and tiny additional bands seen between the two split Dirac cones. These bands shifted from the Fermi level are consistent with the electronic structure of the freestanding trilayer graphene.

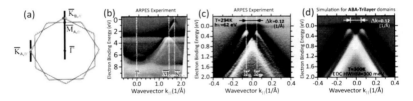

Figure 2.12 Determination of the stacking order in the trilayer graphene grown on SiC(001) from ARPES data. (a) Effective surface Brillouin zone corresponding to four rotated domain variants. (b, c) Dispersion of the π-band in the graphene along the directions indicated in (a). (d) Simulations for ABA-stacked trilayer graphene with a high density of nanodomain boundaries. Reprinted with permission from Ref. [92]. Copyright (2015) American Chemical Society.

To extract information about the stacking order of the nanostructured trilayer graphene on cubic-SiC(001), the fine structure of the multiple bands near the $\overline{K}_A$ and $\overline{K}_{B'}$ points in Fig. 2.12c was simulated for the Bernal- and rhombohedral-stacked trilayer graphene [92]. The simulations were performed taking into account the main rotational variants of the trilayer using the tight-binding (TB) model [120]. Very good correlation between the measured ARPES dispersions (Fig. 2.12c) and simulations (Fig. 2.12d) was achieved for the FWHM of π-bands of 600 meV and for an initial band structure corresponding to the ABA trilayer. Therefore, it was concluded that the stacking order of the trilayer graphene on the cubic-SiC(001) is of ABA type. The FWHM = 600 meV is three times larger than the FWHM measured by ARPES for high-quality single-layer graphene on α-SiC, Ir, or Au [121–123]. Such significantly enhanced broadening can be ascribed either to geometric contributions from a minority of rotational variants distinct from the nanodomains rotated by ±13.5° from the <110> directions or to quantum scattering of quasi particles on structural imperfections. Both origins of the broadening are associated with the formation of the rotated domain network on SiC(001).

2.3.3 Influence of the SiC(001)-c(2×2) Atomic Structure on the Graphene Nanodomain Network

Figure 2.13a shows a schematic model of the graphene overlayer on SiC(001) rotated by 13.5° clockwise relative to the [110] direction. The atomic lattices of the rotated graphene domain and the substrate coincide only in some sites for this misalignment angle, which can lead to a weak interaction between the substrate and overlayer. However, additional points of coincidence can appear due to random distortions of the carbon bond lengths (Fig. 2.9). One can see from Fig. 2.13a that the points of coincidence of the SiC and graphene lattices are almost aligned with the [110] direction of the SiC crystal lattice, corresponding to the direction of atomic chains on the SiC(001)-c(2×2) reconstructed surface (Fig. 2.13b). Most probably, these carbon chains determine the orientation of the domain boundaries in the top layer of the nanostructured few-layer graphene. The coincidence of the directions of carbon atomic chains and graphene domain boundaries is stressed by Fig. 2.13c. One can suppose that the overlayer follows the nanometer-scale morphology of the SiC(001)-c(2×2) surface during the few-layer graphene growth. This is similar to graphene growth on α-SiC [41, 48, 74, 75]. However, because of the different symmetries of the 3C-SiC(001) and honeycomb graphene lattices, they cannot match each other as closely as graphene and α-SiC. This reduces the interaction between the graphene layer and the SiC(001) substrate and leads to the absence of a reactive buffer layer on SiC(001), unlike on α-SiC [41, 48, 74, 75]. It can be supposed that the carbon chains on SiC(001)-c(2×2) or other surface defects (e.g., steps) may be an obstacle for the growth of larger (micrometer-scale) graphene domains on SiC(001), thus defining the orientation of boundaries between neighboring graphene domains. Therefore, controlling the density and orientation of the defects on SiC(001)-c(2×2) could allow the average size of the graphene domains and their orientation to be tuned. This can open a way for synthesis of self-aligned graphene nanoribbons, which can demonstrate unique properties (e.g., open the energy or transport gap in otherwise gapless, semimetallic graphene).

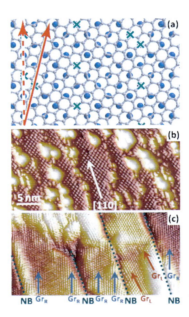

Figure 2.13 (a) Schematic model representing a graphene overlayer rotated by 13.5° clockwise relative to the [110] direction of the underlying C-terminated SiC(001)-c(2×2) surface. The SiC[110] direction is indicated by the dashed arrow and the graphene zigzag direction by the solid arrow. Coinciding lattice points between the graphene layer (gray spheres) and the SiC carbon atoms (blue spheres) are emphasized by green crosses. (b, c) STM images of the SiC(001)-c(2×2) reconstruction (b) and trilayer graphene on SiC(001) (c). The carbon atomic chains on the SiC(001)-c(2×2) surface (b) and the boundaries between nanodomains (NB) in the trilayer graphene (c) are preferentially aligned with the [110] direction. The blue and red arrows in panel (c) stress the rotation of the graphene lattices clockwise (Gr_R) and counterclockwise (Gr_L) relative to the boundaries.

2.4 Nanodomains with Self-Aligned Boundaries on Vicinal SiC(001)/Si(001) Wafers

2.4.1 LEEM and Raman Studies of Graphene/ SiC(001)/4°-off Si(001)

Synthesis of EG on 3C-SiC(001) thin films grown on vicinal Si substrates with a miscut of 4° toward <011> was reported by Ouerghi et al. [91]. Based on LEED, NEXAFS, and Raman spectroscopy data, the authors claimed the fabrication of high-quality graphene without APD boundaries on cubic-SiC(001)/off-axis Si(001). The LEEM data [91] support the synthesis of wafer-scale graphene with two preferential graphene lattice orientations on the vicinal SiC(001) substrates with such miscut angle. However, the Raman data suggest a large number of defects on the graphene synthesized on SiC(001)/4°-off Si(001) wafers.

Figure 2.14 provides typical micro-Raman spectra taken from monolayer and bilayer graphene grown on the vicinal SiC(001) substrate (Fig. 2.14a) and information about the spatial distribution of the 2D peak width (Figs. 2.14b and 2.14d) and 2D peak position (Figs. 2.14c and 2.14e). The Raman spectra reveal three peaks located at 1350, 1594, and 2718 cm^{-1}, which are attributed to the D, G, and 2D bands, respectively. The presence of single G and 2D bands proves the sp^2 reorganization of the topmost layers. In some sample areas the FWHM of the 2D peak is about 60 cm^{-1} (Fig. 2.14a, blue curve), but in the majority of the sample areas, it is about 90 cm^{-1} (Fig. 2.14d). The observed large width of the 2D band and high intensity of the D line were ascribed to a structural disorder and a high density of defects (domain boundaries, vacancies, and distortions). A 14 cm^{-1} blue shift of the G band and a 38 cm^{-1} blue shift of the 2D band (Fig. 2.12e), relative to the values known for the exfoliated graphene, were also observed. That was explained by a compressive strain of the graphene overlayer during the cool down procedure. The Raman and LEED data [91] also suggest that more homogeneous graphene layers on off-axis SiC(001) samples can be obtained using a higher annealing temperature in an argon atmosphere.

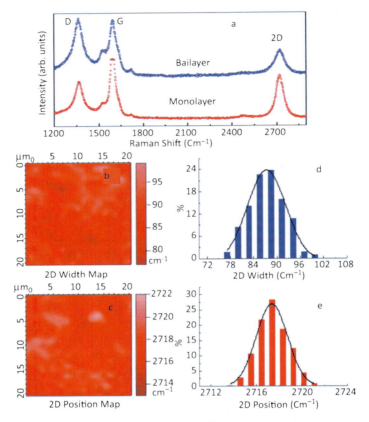

Figure 2.14 (a) Raman spectra taken on the epitaxial graphene layers (red, bilayer; blue, monolayer), (b) false color 2D peak position cartography, (c) false color 2D peak FWHM cartography, (d) distribution of the 2D peak position, and (e) 2D peak FWHM; the bar height indicates the percentage of the total points that had the position (respectively, FWHM) in the 1 cm^{-1} (respectively 2 cm^{-1}) range around the bar. Reprinted with permission from Ref. [91]. Copyright 2012, AIP Publishing LLC.

2.4.2 Atomic and Electronic Structure of the Trilayer Graphene Synthesized on SiC(001)/2°-off Si(001)

In Ref. [92] uniform trilayer graphene was fabricated on vicinal SiC(001)/Si(001) wafers with a miscut of 2°. LEEM, micro-LEED, Raman spectroscopy, NEXAFS, and ARPES studies brought the experimental data similar to that presented earlier for the on-axis SiC(001) sample. Figure 2.15a–c shows bright-field (BF) and tilted BF LEEM micrographs and the corresponding micro-LEED patterns taken from graphene grown on the vicinal SiC(001) substrate. One can see that the UHV-synthesized graphene on the off-axis SiC(001) sample contains APD boundaries (typical APD size for this sample is 0.5–1 μm), which is in contrast with the conclusions of Ref. [91]. Figure 2.15d shows the electron reflectivity spectra acquired in a 7 eV energy window for different sample regions. Three minima are observed across all the curves, which can be attributed to the presence of three graphene monolayers across the whole SiC surface. A micro-LEED pattern taken from a 5 × 5 μm^2 surface area (the inset in Fig. 2.15a) also revealed 12 double-split diffraction spots originating from nanodomains in the graphene trilayer, which have four preferential lattice orientations. The individual micrometer-sized areas, separated by the APD boundaries, usually contain systems of nanodomains with two preferential lattice orientations rotated relative to one another by approximately 27° (Fig. 2.8). The microdomains are contrasted as bright and dark regions in the tilted BF LEEM micrographs. The ARPES experiments (Fig. 2.15e–g) demonstrate the typical spectrum of the trilayer ABA-graphene [92] with intense linear dispersions crossing E_F and less intense split bands shifted from the Fermi level. Angular dependence of the NEXAFS spectra measured on the vicinal sample (Fig. 2.15h) is also in agreement with that taken from the on-axis SiC(001) samples (Fig. 2.4c). One can only note a slightly higher intensity of the peak related to 1s → σ* transition at the photon incidence angle of 85° for the vicinal sample, which can be related to the smaller size of the antiphase domains and higher density of defects on the vicinal sample studied.

STM studies revealed that the microdomains contain systems of nanodomains that are preferentially elongated in one direction, which is very close to the [110] direction and the step direction of the vicinal SiC(001) sample before graphene synthesis. This is illustrated by Figs. 2.16a and 2.16b. Remarkably, it was found that the direction of the NBs in nanostructured graphene on the 2°-off SiC(001)/Si(001) sample was not changing even at the APD boundaries and intrinsic surface defects [92].

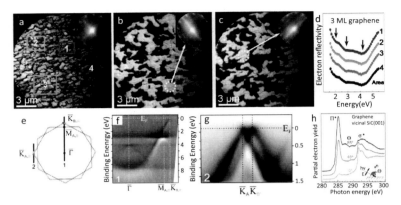

Figure 2.15 LEEM, micro-LEED, ARPES, and NEXAFS characterizations of trilayer graphene synthesized on a vicinal SiC(001) substrate. (a) 15 × 15 μm² BF LEEM image recorded with an electron energy of 54 eV. The dark lines in the image are the APD boundaries. Inset: micro-LEED pattern (E = 45 eV) from a 5 × 5 μm² surface area. The probing area covers nanodomain systems with the lattices rotated by ±13.5° relative to [110] and [1–10] directions. All 24 graphene spots are present in the diffraction pattern. The (1,0) and (0,1) SiC substrate spots are clearly seen. (b, c) Tilted BF LEEM images (electron energy 6 eV) acquired by deflecting the incident (0,0) electron beam. The reverse contrast in the images is due to the presence of two orthogonal nanodomain families. Insets: micro-LEED patterns taken from the bright areas on the corresponding LEEM images. (d) The electron reflectivity curves acquired from the areas indicated in panel (a). All curves demonstrate three consistent minima (indicated by arrows), proving the uniform trilayer thickness of the graphene overlayer. (e) Effective surface Brillouin zone. (f, g) Dispersion of the π-band measured by ARPES along directions 1 and 2 indicated in panel (e). (h) NEXAFS spectra taken at three different incidence angles of the linearly polarized photons.

Figure 2.16c shows an atomically resolved STM image containing three nanometer-scale domains connected to each other through the NBs. Detailed analysis of the high-resolution STM images measured near the boundaries shows that, in most cases, they are rotated by 3.5° relative to the [110] direction, as depicted in the schematic shown on Fig. 2.16e. Since the graphene nanodomain lattices are rotated by ±13.5° from the same [110] direction, they are asymmetrically rotated relative to the boundaries (Fig. 2.16c). The lattices in neighboring domains are rotated by 10° anticlockwise (Gr$_L$) and 17° clockwise (Gr$_R$) relative to the NB. As Fig. 2.16e illustrates, this asymmetry near the NB leads to the formation of a periodic structure along the boundaries, with a period of 1.37 nm. The periodic structure consists of distorted heptagons and pentagons (Fig. 2.16e), which is consistent with the atomically resolved STM image measured at the NB (Fig. 2.16d).

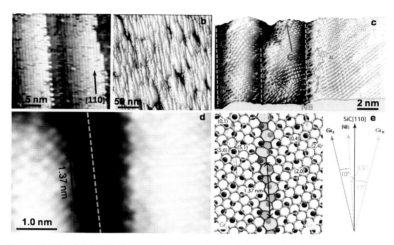

Figure 2.16 (a) STM characterization of the vicinal SiC(001)3×2 surface. The image demonstrates that the step direction is close to the [110] direction of the SiC crystal lattice. The image was measured at $U = -2.3$ V and $I = 80$ pA. (b) Large-area STM image of the self-aligned graphene nanoribbons synthesized on the vicinal SiC(001). The domain boundaries are preferentially aligned with the [110] direction. (c, d) Atomically resolved STM images of graphene nanoribbons showing the system of domains rotated 17° clockwise (Gr_R) and 10° counterclockwise (Gr_L) relative to the NB, which is rotated 3.5° anticlockwise from the [110] direction (c) and the atomic structure of the NB (d). The images were measured at $U = -100$ mV and $I = 68$ pA. (e) Schematic model of the NB for the asymmetrically rotated nanodomains in panels (c) and (d). For the angles shown, a periodic structure of distorted pentagons and heptagons is formed. Reprinted with permission from Ref. [92]. Copyright (2015) American Chemical Society.

2.4.3 Transport Gap Opening in Nanostructured Trilayer Graphene with Self-Aligned Domain Boundaries

Recent theoretical works [124] showed that graphene NBs with a periodic atomic structure along their lengths could perfectly reflect charge carriers over a large range of energies. Harnessing this would provide a new way to control the charge carriers without the need to introduce an energy bandgap. The main challenge is to produce self-aligned nanodomains with periodic NBs on a semiconducting substrate compatible with existing processes, which, apparently, can be overcome using the SiC epilayers grown on vicinal Si(001) wafers.

Figure 2.17a shows a schematic drawing of a typical graphene nanogap device fabricated by electron beam lithography on graphene/SiC(001)/2°-off Si(001) and used for the transport measurements reported in Ref. [92]. In the experiments the bias voltage was applied perpendicular to the NBs to measure the local transport properties due to the formation of the system of asymmetrically rotated domains shown (Fig. 2.16). It was suggested by Yazyev and Louie [124] that a charge transport gap of $E_g = \hbar v_F \dfrac{2\pi}{3d} \approx \dfrac{1.38}{d(nm)}(eV)$ can be formed by a nonsymmetric NB associated with a lattice mismatch at the boundary line, where $\hbar$ is the reduced Planck's constant, v_F is the Fermi velocity, and d is the periodicity along the NB. As indicated in Figs. 2,16d and 2.16e, the asymmetric rotation of the graphene lattices in the neighboring domains relative to the NB leads to a 1.37 nm periodicity along the NB. According to the theory [124], this periodicity should produce a transport gap of approximately 1.0 eV, which is consistent with the transport measurements presented in Fig. 2.17.

Figures 2.17b and 2.17c show the $I–V$ curves measured using the nanogap device at different temperatures. The transport gap is clearly observed below 100 K but disappears at temperatures above 150 K. To define the exact values of the transport gaps observed for trilayer graphene grown on 2°-off SiC(001), the corresponding dI/dV curves are plotted in Fig. 2.17d for temperatures below 150 K. For a small bias voltage, any reasonable current signal cannot be detected and the corresponding dI/dV is around 0.01 µS, indicating the existence of a transport gap. The transport gap derived from the dI/dV curve is approximately 1.3 eV at 10 K. Remarkably, the transport gap is approximately the same at 50 K and 10 K but substantially lower (0.4 eV) at 100 K. The conductivity of the device is only 10^{-2} µS at bias voltages smaller than the transport gap, but this increases to 10^2 µS when the bias voltage is larger than the transport gap, which gives a high on-off current ratio of 10^4. Note that in the nanogap contact devices the NBs are uniform and directed along the step direction of the vicinal SiC(001) (Fig. 2.16), which gives this system the potential for high-density memory applications.

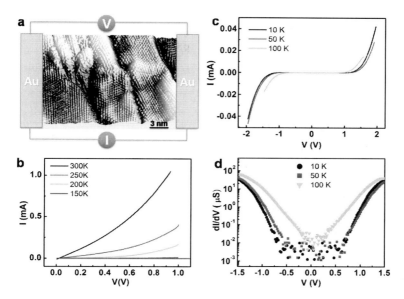

Figure 2.17 Electrical detection of the opening of a transport gap in trilayer graphene on a vicinal SiC substrate. (a) Schematic drawing of the nanogap device. (b–d) I–V curves measured at different temperatures with current across the self-aligned NBs. (b) I–V curves measured at 150 K, 200 K, 250 K, and 300 K. (c) I–V curves measured at 10 K, 50 K, and 100 K. (d) Corresponding dI/dV curves for temperatures below 150 K. Reprinted with permission from Ref. [92]. Copyright (2015) American Chemical Society.

The disappearance of the transport gap at temperatures above 150 K in the experiments shown in Fig. 2.17 can be related to the presence of defects in the graphene trilayer, which can remarkably modify the transport properties of graphite and graphene [125]. For example, transport properties of the nanostructured graphene with self-aligned boundaries can be modified by interstitials, which may be frozen under 100 K and migrate at higher temperatures (100–300 K) [126].

To understand the effect of the NBs on the transport properties in the nanostructured trilayer graphene grown on the vicinal SiC(001) substrate, electric current across the NBs was simulated as a function of gate voltage [92]. In the calculations, the gate voltage was adjusted through the on-site energy. A nonequilibrium Green function was used, and the current was calculated with Landauer–Keldysh formalism [127, 128]. The unit cell used for the calculations is shown in Fig. 2.18a, and the results of the simulations are summarized in Fig. 2.18b. There is a plateau region where electrons cannot transport through the boundary, indicating that a gap does open with the current driven across the self-aligned NBs and can be tuned with the gate voltage.

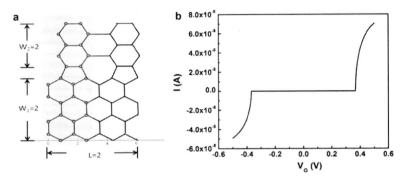

Figure 2.18 Current as a function of gate voltage. (a) Schematic drawing of the model used, where L is the length of the NB, W_1 is the width of the armchair structure, and W_2 is the width of the zigzag structure. (b) Current as a function of gate voltage calculated from first-principles simulation. Reprinted with permission from Ref. [92]. Copyright (2015) American Chemical Society.

The I–V curves were also measured from graphene/SiC(001)/2°-off Si(001) at 10 K, with the current applied along the boundaries (Fig. 2.19). No transport gap is observed, and the I–V curve displays nonlinear behavior. This indicates that the observed charge transport gap for current across the NBs is mainly due to the self-aligned periodic NBs, which can reflect charge carriers over a range of energies. In addition to this, photoemission experiments on the vicinal SiC(001) samples were conducted at temperatures between 50 K and 300 K. The experimental data undoubtedly showed the unchanged position of the valence-band edge with a decreasing temperature, proving that the observed transport gap is related to a specific atomic structure of the NBs rather than to a bandgap opening in the nanostructured graphene during the sample cooling. At the same time, the secondary electron cutoff measurements revealed a change of the work function at temperatures below 150 K [92], which are in agreement with the transport measurements shown in Fig. 2.17.

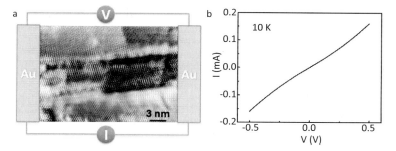

Figure 2.19 *I–V* measurements with the current applied along the NBs. (a) Schematic of the nanogap device and (b) *I–V* curve measured at 10 K along NBs (reprinted with permission from Ref. [92]. Copyright (2015) American Chemical Society).

2.5 Conclusions

The data presented in this chapter demonstrate that continuous and uniform few-layer graphene can be synthesized in UHV on low-cost, technologically relevant SiC(001)/Si(001) wafers. The continuity of the graphene overlayer is not broken by the antiphase domain boundaries, which are typical defects of the SiC(001) thin films grown on Si(001). The results obtained by various techniques prove very weak interaction of the graphene overlayer with the SiC substrate. The synthesized few-layer graphene demonstrates the properties of the quasi-freestanding graphene (atomic-scale rippling, carbon–carbon bond distortions, linear dispersions with the Dirac points at the Fermi level, high flexibility of the topmost layer, etc.). The continuous few-layer graphene coverage on SiC(001) consists of rotated nanodomains with several preferential lattice orientations. Such nanodomain system with a periodic structure along the boundaries can produce a charge transport gap in the gapless, semimetallic graphene. Therefore, a possibility to synthesize graphene with self-aligned NBs on SiC(001) substrates can open up opportunities for a wide range of new applications. Recent studies conducted on graphene/SiC(001)/2°-off Si(001) samples demonstrate a possibility to synthesize nanodomains with self-aligned boundaries. This system has been utilized to achieve a current on-off ratio of 10^4 by opening a transport gap in the nanostructured trilayer graphene.

Acknowledgments

This work was partially supported by the Russian Academy of Sciences, Russian Foundation for Basic Research (grants 14-02-00949, 14-02-01234, 17-02-01139, and 17-02-01291), and Marie Curie IIF grant within the 7th European Community Framework Programme. We thank our colleagues S. V. Babenkov, H.-C. Wu, S. N. Molotkov, D. Marchenko, A. Varykhalov, A. A. Zakharov, B. E. Murphy, S. A. Krasnikov, I. V. Shvets, S. L. Molodtsov, and J. Viefhaus for fruitful discussions and help in preparation of this manuscript.

References

1. Novoselov, K. S., Geim, A. K., Morozov, S. V., Jiang, D., Zhang, Y., Dubonos, S. V., Grigorieva, I. V., and Firsov, A. A., Electric field effect in atomically thin carbon films, *Science*, **306**, 666–669 (2004).

2. Geim, A. K., and Novoselov, K. S., The rise of graphene, *Nat. Mater.*, **6**, 183–191 (2007).

3. Heersche, H. B., Jarillo-Herrero, P., Oostinga, J. B., Vandersypen, L. M. K., and Morpurgo, A. F., Bipolar supercurrent in graphene, *Nature*, **446**, 56–59 (2007).

4. Zhang, Y., Jiang, Z., Small, J. P., Purewal, M. S., Tan, Y.-W., Fazlollahi, M., Chudow, J. D., Jaszczak, J. A., Stormer, H. L., and Kim, P., Landau-level splitting in graphene in high magnetic fields, *Phys. Rev. Lett.*, **96**, 136806(4) (2006).

5. Zhang, Y., Tan, Y.-W., Stormer, H. L., and Kim, P., Experimental observation of the quantum Hall effect and Berry's phase in graphene, *Nature*, **438**, 201–204 (2005).

6. Novoselov, K. S., Geim, A. K., Morozov, S. V., Jiang, D., Katsnelson, M. I., Grigorieva, I. V., Dubonos, S. V., and Firsov, A. A., Two-dimensional gas of massless Dirac fermions in graphene, *Nature*, **438**, 197–200 (2005).

7. Novoselov, K. S., Nobel lecture: graphene; materials in the Flatland, *Rev. Mod. Phys.*, **83**, 837–849 (2011).

8. Novoselov, K. S., Falko, V. I., Colombo, L., Gellert, P. R., Schwab, M. G., and Kim, K., A roadmap for graphene, *Nature*, **490**, 192–200 (2012).

9. Zhou, S. Y., Gweon, G.-H., Fedorov, A. V., First, P. N., de Heer, W. A., Lee, D.-H., Guinea, F., Castro Neto, A. H., and Lanzara, A., Substrate-induced bandgap opening in epitaxial graphene, *Nat. Mater.*, **6**, 770–775 (2007).

10. Chen, Z., Lin, Y.-M., Rooks, M. J., and Avouris, P., Graphene nano-ribbon electronics, *Physica E*, **40**, 228–232 (2007).

11. de Heer, W. A., Berger, C., Wu, X., First, P. N., Conrad, E. H., Li, X., Li, T., Sprinkle, M., Hass, J., Sadowski, M. L., Potemski, M., and Martinez, G., Epitaxial graphene, *Solid State Commun.*, **143**, 92–100 (2007).

12. Geim, A. K., Review. Graphene: status and prospects, *Science*, **324**, 1530–1534 (2009).

13. Berger, C., Song, Z., Li, X., Wu, X., Brown, N., Naud, C., Mayou, D., Li, T., Hass, J., Marchenkov, A. N., Conrad, E. H., First, P. N., and de Heer, W. A., Electronic confinement and coherence in patterned epitaxial graphene, *Science*, **312**, 1191–1196 (2006).

14. Novoselov, K. S., Jiang, D., Schedin, F., Booth, T. J., Khotkevich, V. V., Morozov, S.V., and Geim A. K., Two-dimensional atomic crystals, *Proc. Natl. Acad. Sci. U S A*, **102**, 10451–10453 (2005).

15. Stankovich, S., Dikin, D. A., Dommett, G. H. B., Kohlhaas, K. M., Zimney, E. J., Stach, E. A., Piner, R. D., Nguyen, SB. T., and Ruoff, R. S., Graphene-based composite materials, *Nature*, **442**, 282–286 (2006).

16. Hernandez, Y., Nicolosi, V., Lotya, M., Blighe, F. M., Sun, Z., De, S., McGovern, I. T., Holland, B., Byrne, M., Gun'Ko, Y. K., Boland, J. J., Niraj, P., Duesberg, G., Krishnamurthy, S., Goodhue, R., Hutchison, J., Scardaci, V., Ferrari, A. C., and Coleman J. N., High-yield production of graphene by liquid-phase exfoliation of graphite, *Nat. Nanotechnol.*, **3**, 563–568 (2008).

17. Blake, P., Brimicombe, P. D., Nair, R. R., Booth, T. J., Jiang, D., Schedin, F., Ponomarenko, L. A., Morozov, S. V., Gleeson, H. F., Hill, E. W., Geim, A. K., and Novoselov, K. S., Graphene-based liquid crystal device, *Nano Lett.*, **8**, 1704–1708 (2008).

18. Hagstrom, S., Lyon, H. B., Somorjai, G. A., Surface structures on the clean platinum (100) surface, *Phys. Rev. Lett.*, **15**, 491–493 (1965).

19. Lyon, H. B., and Somorjai, G. A., Low energy electron diffraction study of the clean (100), (111), and (110) faces of platinum, *J. Chem. Phys.*, **46**, 2539–2550 (1967).

20. Morgan, A. E., and Somorjai, G. A., LEED studies of gas adsorption on the platinum (100) single crystal surface, *Surf. Sci.*, **12**, 405–425 (1968).

21. May, J. W., Platinum surface LEED rings, *Surf. Sci.*, **17**, 267–270 (1969).

22. Grant, J. T., and Haas, T. W., A study of Ru(0001) and Rh(111) surfaces using LEED and Auger electron spectroscopy, *Surf. Sci.*, **21**, 76–85 (1970).

23. Schlögl, R., Ertl, G., Knözinger, H., Schüth, F., and Weitkamp, J. (eds.), *Handbook of Heterogeneous Catalysis*, Vol. 1, Wiley-VCH, Weinheim, p. 357 (2008).

24. Moulijn, J. A., van Diepen, A. E., Kapteijn F., Ertl, G., Knözinger, H., Schüth, F., and Weitkamp, J. (eds.), *Handbook of Heterogeneous Catalysis*, Vol. 4, Wiley-VCH, Weinheim, p. 1829 (2008).

25. Batzill, M., The surface science of graphene: metal interfaces, CVD synthesis, nanoribbons, chemical modifications, and defects, *Surf. Sci. Rep.*, **67**, 83–115 (2012).

26. Wintterlin, J., and Bocquet M.-L., Graphene on metal surfaces, *Surf. Sci.*, **603**, 1841–1852 (2009).

27. Grüneis, A., and Vyalikh, D. V., Tunable hybridization between electronic states of graphene and a metal surface, *Phys. Rev. B*, **77**, 193401(4) (2008).

28. Park, S., and Ruoff, R. S., Chemical methods for the production of graphenes, *Nat. Nanotechnol.*, **4**, 217–224 (2009).

29. Michon, A., Vézian, S., Ouerghi, A., Zielinski, M., Chassagne, T., and Portail, M., Direct growth of few-layer graphene on 6H-SiC and 3C-SiC/Si via propane chemical vapor deposition, *Appl. Phys. Lett.*, **97**, 171909 (2010).

30. Wang, J. J., Zhu, M. Y., Outlaw, R. A., Zhao, X., Manos, D. M., Holloway, B. C., and Mammana, V. P., Free-standing subnanometer graphite sheets, *Appl. Phys. Lett.*, **85**, 1265–1267 (2004).

31. Wang, J., Zhu, M., Outlaw, R. A., Zhao, X., Manos, D. M., and Holloway, B. C., Synthesis of carbon nanosheets by inductively coupled radio-frequency plasma enhanced chemical vapor deposition, *Carbon*, **42**, 2867–2872 (2004).

32. Malesevic, A., Vitchev, R., Schouteden, K., Volodin, A., Zhang, L., Van Tendeloo, G., Vanhulsel, A., and Van Haesendonck, C., Synthesis of few-layer graphene via microwave plasma-enhanced chemical vapour deposition, *Nanotechnology*, **19**, 305604(6) (2008).

33. Vitchev, R., Malesevic, A., Petrov, R. H., Kemps, R., Mertens, M., Vanhulsel A., and Van Haesendonck, C., Initial stages of few-layer graphene growth by microwave plasma-enhanced chemical vapour deposition, *Nanotechnology*, **21**, 095602(7) (2010).

34. Gomez-Navarro, C., Burghard, M., and Kern, K., Elastic properties of chemically derived single graphene sheets, *Nano Lett.*, **8**, 2045–2049 (2008).

35. Verdejo, R., Barroso-Bujans, F., Rodriguez-Perez, M. A., Antonio de Saja, J., and Lopez-Manchado, M. A., Functionalized graphene sheet filled silicone foam nanocomposites, *J. Mater. Chem.*, **18**, 2221–2226 (2008).

36. Schniepp, H. C., Li, J.-L., McAllister, M. J., Sai, H., Herrera-Alonso, M., Adamson, D. H., Prud'homme, R. K., Car, R., Saville, D. A., and Aksay, I. A., Functionalized single graphene sheets derived from splitting graphite oxide, *J. Phys. Chem. B*, **110**, 8535–8539 (2006).

37. Gilje, S., Han, S., Wang, M., Wang, K. L., and Kaner, R. B., A chemical route to graphene for device applications, *Nano Lett.*, **7**, 3394–3398 (2007).

38. Rollings, E., Gweon, G. H., Zhou, S. Y., Mun, B. S., McChesney, J. L., Hussain, B. S., Fedorov, A. V., First, P. N., de Heer, W. A., and Lanzara, A., Synthesis and characterization of atomically thin graphite films on a silicon carbide substrate, *J. Phys. Chem. Solids*, **67**, 2172–2177 (2006).

39. Hass, J., Feng, R., Li, T., Li, X., Zong, Z., de Heer, W. A., First, P. N., Conrad, E. H., Jeffrey, C. A., and Berger, C., Highly ordered graphene for two dimensional electronics, *Appl. Phys. Lett.*, **89**, 143106(3) (2006).

40. Virojanadara, C., Syväjarvi, M., Yakimova, R., Johansson, L. I., Zakharov, A. A., and Balasubramanian, T., Homogeneous large-area graphene layer growth on 6H-SiC(0001), *Phys. Rev. B*, **78**, 245403(6) (2008).

41. Emtsev, K. V., Bostwick, A., Horn, K., Jobst, J., Kellogg, G. L., Ley, L., McChesney, J. L., Ohta, T., Reshanov, S. A., Röhrl, J., Rotenberg, E., Schmid, A. K., Waldmann, D., Weber, H. B., and Seyller, T., Towards wafer-size graphene layers by atmospheric pressure graphitization of silicon carbide, *Nat. Mater.*, **8**, 203–207 (2009).

42. van Bommel, A. J., Crombeen J. E., and van Tooren, A., LEED and Auger electron observations of the SiC(0001) surface, *Surf. Sci.*, **48**, 463–472 (1975).

43. Berger, C., Song, Z., Li, T., Li, X., Ogbazghi, A. Y., Feng, R., Dai, Z., Marchenkov, A. N., Conrad, E. H., First, P. N., and de Heer, W. A., et al., Ultrathin epitaxial graphite: 2D electron gas properties and a route toward graphene-based nanoelectronics, *J. Phys. Chem. B*, **108**, 19912–19916 (2004).

44. Forbeaux, I., Themlin, J.-M., and Debever, J.-M., Heteroepitaxial graphite on 6H–SiC(0001): interface formation through conduction-band electronic structure, *Phys. Rev. B*, **58**, 16396–16406 (1998).

45. Ohta, T., Bostwick, A., Seyller, T., Horn, K., and Rotenberg, E., Controlling the electronic structure of bilayer graphene, *Science*, **313**, 951–954 (2006).

46. Riedl, C., Starke, U., Bernhardt, J., Franke, M., and Heinz, K., Structural properties of the graphene-SiC(0001) interface as a key for the preparation of homogeneous large-terrace graphene surfaces, *Phys. Rev. B*, **76**, 245406 (2007).

47. Lin, Y.-M., Dimitrakopoulos, C., Jenkins, K. A, Farmer, D. B., Chiu, H.-Y., Grill, A., and Avouris, P., 100-GHz transistors from wafer-scale epitaxial graphene, *Science*, **327**, 662 (2010).

48. Sprinkle, M., Siegel, D., Hu, Y., Hicks, J., Tejeda, A., Taleb-Ibrahimi, A., Le Fèvre, P., Bertran, F., Vizzini, S., Enriquez, H., Chiang, S., Soukiassian, P., Berger, C., de Heer, W. A., Lanzara, A., and Conrad, E. H., First direct observation of a nearly ideal graphene band structure, *Phys. Rev. Lett.*, **103**, 226803 (2009).

49. Nishino, S., Powell, J. A., and Will, H. A., Production of large-area single-crystal wafers of cubic SiC for semiconductor devices, *Appl. Phys. Lett.*, **42**, 460–462 (1983).

50. Feng, Z. C., Mascarenhas, A. J., Choyke, W. J., and Powell, J. A. (1988). Raman scattering studies of chemical-vapor-deposited cubic SiC films of (100) Si, *J. Appl. Phys.*, **64**, 3176–3186 (1988).

51. Shigeta, M., Fujii, Y., Furukawa, K., Suzuki, A., and Nakajima, S., Chemical vapor deposition of single-crystal films of cubic SiC on patterned Si substrates, *Appl. Phys. Lett.*, **55**, 1522–1524 (1989).

52. Golecki, I., Reidinger, F., and Marti, J., Single-crystalline, epitaxial cubic SiC films grown on (100) Si at 750 °C by chemical vapor deposition, *Appl. Phys. Lett.*, **60**, 1703–1705 (1992).

53. Hoechst, H., Tang, M., Johnson, C., Meese, J. M., Zajac, G. W., and Fleisch, T. H., The electronic structure of cubic SiC grown by chemical vapor deposition on Si(100), *J. Vac. Sci. Technol. A*, **5**, 1640–1643 (1987).

54. Aristov, V. Y., β-SiC (100) surface: atomic structures and electronic properties, *Phys.: Uspekhi (Rev. Topical Problems)*, **44**, 761–783 (2001).

55. Soukiassian, P. G., and Enriquez, H. B., Atomic scale control and understanding of cubic silicon carbide surface reconstructions, nanostructures and nanochemistry, *J. Phys.: Condens. Matter.*, **16**, S1611–S1658 (2004).

56. Aristov, V. Y., Urbanik, G., Kummer, K., Vyalikh, D. V., Molodtsova, O. V., Preobrajenski, A. B., Zakharov, A. A., Hess, C., Haenke, T., Buechner, B., Vobornik, I., Fujii, J., Panaccione, G., Ossipyan, Y. A., and Knupfer, M.,

Graphene synthesis on cubic SiC/Si wafers. Perspectives for mass production of graphene-based electronic devices, *Nano Lett.*, **10**, 992–995 (2010).

57. Suemitsu, M., and Fukidome, H., Epitaxial graphene on silicon substrates, *J. Phys. D: Appl. Phys.*, **43**, 374012 (2010).

58. Fukidome, H., Miyamoto, Y., Handa, H., Saito, H., and Suemitsu M., Epitaxial growth processes of graphene on silicon substrates, *Jpn. J. Appl. Phys.*, **49**, 01AH03-4 (2010).

59. Otsuji, T., Tombet, S. A. B., Satou, A., Fukidome, H., Suemitsu, M., and Sano, E., Graphene-based devices in terahertz science and technology, *J. Phys. D: Appl. Phys.*, **45**, 303001 (2012).

60. Ouerghi, A., Kahouli, A., Lucot, D., Portail, M., Travers, L., Gierak, J., Penuelas, J., Jegou, P., Shukla, A., Chassagne, T., and Zielinski, M., Epitaxial graphene on cubic SiC(111)/Si(111) substrate, *Appl. Phys. Lett.*, **96**, 191910 (2010).

61. Coletti, C., Emtsev, K. V., Zakharov, A. A., Ouisse, T., Chaussende, D., and Starke, U., Large area quasi-free standing monolayer graphene on 3C-SiC(111), *Appl. Phys. Lett.*, **99**, 081904 (2011).

62. Portail, M., Michon, A., Vezian, S., Lefebvre, D., Chenot, S., Roudon, E., Zielinski, M., Chassagne, T., Tiberj, A., Camassel, J., and Cordier, Y., Growth mode and electric properties of graphene and graphitic phase grown by argon–propaneassisted CVD on 3C–SiC/Si and 6H–SiC, *J. Cryst. Growth*, **349**, 27–35 (2012).

63. Abe, S., Handa, H., Takahashi, R., Imaizumi, K., Fukidome, H., and Suemitsu, M., Surface chemistry involved in epitaxy of graphene on 3C-SiC(111)/Si(111), *Nanoscale Res. Lett.*, **5**, 1888–1891 (2010).

64. Ouerghi, A., Belkhou, R., Marangolo, M., Silly, M. G., El Moussaoui, S., Eddrief, M., Largeau, L., Portail, M., and Sirotti, F., Structural coherency of epitaxial graphene on 3C-SiC(111) epilayers on Si(111), *Appl. Phys. Lett.*, **97**, 161905 (2010).

65. Ouerghi, A., Marangolo, M., Belkhou, R., El Moussaoui, S., Silly, M. G., Eddrief, M., Largeau, L., Portail, M., Fain, B., and Sirotti, F., Epitaxial graphene on 3C-SiC(111) pseudosubstrate: structural and electronic properties, *Phys. Rev. B*, **82**, 125445 (2010).

66. Takahashi, R., Handa, H., Abe, S., Imaizumi, K., Fukidome, H., Yoshigoe, A., Teraoka Y., and Suemitsu, M., Low-energy-electron-diffraction and X-ray-phototelectron-spectroscopy studies of graphitization of 3C-SiC(111) thin film on Si(111) substrate, *Jpn. J. Appl. Phys.*, **50**, 070103 (2011).

67. Aryal, H. R., Fujita, K., Banno, K., and Egawa, T., Epitaxial graphene on Si(111) substrate grown by annealing 3C-SiC/carbonized silicon, *Jpn. J. Appl. Phys.*, **51**, 01AH05 (2012).

68. Miyamoto, Y., Handa, H., Saito, E., Konno, A., Narita, Y., Suemitsu, M., Fukidome, H., Ito, T., Yasui, K., Nakazawa, H., and Endoh, T., Raman-scattering spectroscopy of epitaxial graphene formed on SiC Film on Si substrate, *e-J. Surf. Sci. Nanotechnol.*, **7**, 107–109 (2009).

69. Suemitsu, M., Miyamoto, Y., Handa, H., and Konno, A., *e-J. Surf. Sci. Nanotechnol.*, **7**, 311–313 (2009).

70. Fukidome, H., Abe, S., Takahashi, R., Imaizumi, K., Inomata, S., Handa, H., Saito, E., Enta, Y., Yoshigoe, A., Teraoka, Y., Kotsugi, M., Ohkouchi, T., Kinoshita, T., Ito, S., and Suemitsu, M., Controls over structural and electronic properties of epitaxial graphene on silicon using surface termination of 3C-SiC(111)/Si, *Appl. Phys. Express*, **4**, 115104 (2011).

71. Hsia, B., Ferralis, N., Senesky, D. G., Pisano, A. P., Carraro C., and Maboudian, R., Epitaxial graphene growth on 3C SiC(111)/AlN(0001)/Si(100), *Electrochem. Solid State Lett.*, **14**, K13–K15 (2011).

72. Starke, U., Coletti, C., Emtsev, K., Zakharov, A. A., Ouisse, T., and Chaussende, D., Large area quasi-free standing monolayer graphene on 3C-SiC (111), *Mater. Sci. Forum*, **717–720**, 617–620 (2012).

73. Fukidome, H., Ide, T., Kawai, Y., Shinohara, T., Nagamura, N., Horiba, K., Kotsugi, M., Ohkochi, T., Kinoshita, T., Kumighashira, H., Oshima M., and Suemitsu, M., Microscopically-tuned band structure of epitaxial graphene through Interface and stacking variations using Si substrate microfabrication, *Sci. Rep.*, **4**, 5173–5176 (2014).

74. Emstev, K. V., Speck, F., Seyller, T., Ley, L., and Riley, J. D., Interaction, growth, and ordering of epitaxial graphene on SiC{0001} surfaces: a comparative photoelectron spectroscopy study, *Phys. Rev. B*, **77**, 155303 (2008).

75. Hass, J., de Heer, W. A., and Conrad, E. H., The growth and morphology of epitaxial multilayer graphene, *J. Phys.: Condens. Matter*, **20**, 323202 (2008).

76. Nakao, M., Iikawa, H., Izumi, K., Yokoyama, T., and Kobayashi, S., Challenge to 200 mm 3C-SiC wafers using SOI, *Mater. Sci. Forum*, **483–485**, 205–208 (2005).

77. Ouerghi, A., Ridene, M., Balan, A., Belkhou, R., Barbier, A., Gogneau, N., Portail, M., Michon, A., Latil, S., Jegou, P., and Shukla, A., Sharp interface in epitaxial graphene layers on3C-SiC(100)/Si(100) wafers, *Phys. Rev. B*, **83**, 205429 (2011).

78. Gogneau, N., Balan, A., Ridene, M., Shukla, A., and Ouerghi, A., Control of the degree of surface graphitization on 3C-SiC(100)/Si(100), *Surf. Sci.*, **606**, 217–220 (2012).

79. Ouerghi, A., Balan, A., Castelli, C., Picher, M., Belkhou, R., Eddrief, M., Silly, M. G., Marangolo, M., Shukla, A., and Sirotti, F., Epitaxial graphene on single domain 3C-SiC(100) thin films grown on off-axis Si(100), *Appl. Phys. Lett.*, **101**, 021603 (2012).

80. Huang, H., Wong, S. L., Tin, C., Luo, Z. Q., Shen, Z. X., Chen, W., and Wee, A. T. S., Epitaxial growth and characterization of graphene on free-standing polycrystalline 3C-SiC, *J. Appl. Phys.*, **110**, 014308 (2011).

81. Chaika, A. N., Molodtsova, O. V., Zakharov, A. A., Marchenko, D., Sanchez-Barriga, J., Varykhalov, A., Shvets, I. V., and Aristov, V. Y., Continuous wafer-scale graphene on cubic-SiC(001), *Nano Res.*, **6**, 562–570 (2013).

82. Abe, S., Handa, H., Takahashi, R., Imaizumi, K., Fukidome, H., and Suemitsu, M., Temperature-programmed desorption observation of graphene-on-silicon process, *Jpn. J. Appl. Phys.*, **50**, 070102(5) (2011).

83. Gogneau, N., Balan, A., Ridene, M., Shukla, A., and Ouerghi, A., Control of the degree of surface graphitization on 3C-SiC(100)/Si(100), *Surf. Sci.*, **606**, 217–220 (2012).

84. Chaika, A. N., Molodtsova, O. V., Zakharov, A. A., Marchenko, D., Sánchez-Barriga, J., Varykhalov, A., Shvets, I. V., and Aristov, V. Y., Continuous wafer-scale graphene on cubic-SiC(001), *Nano Res.*, **6**, 562–570 (2013).

85. Chaika, A. N., Molodtsova, O. V., Zakharov, A. A., Marchenko, D., Sánchez-Barriga, J., Varykhalov, A., Babenkov, S.V., Portail, M., Zielinski, M., Murphy, B.E., Krasnikov, S.A., Lübben, O., Shvets, I. V., and Aristov, V. Y., Rotated domain network in graphene on cubic-SiC(001), *Nanotechnology*, **25**, 135605(8) (2014).

86. Velez-Fort, E., Silly, M. G., Belkhou, R., Shukla, A., Sirotti, F., and Ouerghi, A., Edge state in epitaxial nanographene on 3C-SiC(100)/Si(100) substrate, *Appl. Phys. Lett.*, **103**, 083101(4) (2013).

87. Gogneau, N., Ben Gouider Trabelsi, A., Silly, M. G., Ridene, M., Portail, M., Michon, A., Oueslati, M., Belkhou, R., Sirotti, F., and Ouerghi, A., Investigation of structural and electronic properties of epitaxial graphene on 3C–SiC(100)/Si(100) substrates, *Nanotechnol. Sci. Appl.*, **7**, 85–95 (2014).

88. Hens, P., Zakharov, A. A., Iakimov, T., Syväjärvi, M., and Yakimova, R., Large area buffer-free graphene on non-polar (001) cubic silicon carbide, *Carbon*, **80**, 823–829 (2014).

89. Suemitsu, M., Jiao, S., Fukidome, H., Tateno, Y., Makabe, M., and Nakabayashi, T., Epitaxial graphene formation on 3C-SiC/Si thin films, *J. Phys. D: Appl. Phys.*, **47**, 094016(11) (2014).

90. Iacopi, F., Mishra, N., Cunning, B. V., Goding, D., Dimitrijev, S., Brock, R., Dauskardt, R. H., Wood, B., and Boeckl, J., A catalytic alloy approach for graphene on epitaxial SiC on silicon wafers, *J. Mater. Res.*, **30**, 609–616 (2015).

91. Ouerghi, A., Balan, A., Castelli, C., Picher, M., Belkhou, R., Eddrief, M., Silly, M. G., Marangolo, M., Shukla, A., and Sirotti, F., Epitaxial graphene on single domain 3C-SiC(100) thin films grown on off-axis Si(100), *Appl. Phys. Lett.*, **101**, 021603(5) (2012).

92. Wu, H.-C., Chaika, A. N., Huang, T.-W., Syrlybekov, A., Abid, M., Aristov, V. Y., Molodtsova, O. V., Babenkov, S. V., Marchenko, D., Sánchez-Barriga, J., Mandal, P. S., Varykhalov, A. Y., Niu, Y., Murphy, B. E., Krasnikov, S. A., Lübben, O., Wang, J-J., Liu, H., Yang, L., Zhang, H., Abid, M., Janabi, J. T., Molotkov, S. N., Chang, C-R., and Shvets, I. V., Transport gap opening and high on-off current ratio in trilayer graphene with self-aligned nanodomain boundaries, *ACS Nano*, **9**, 8967–8975 (2015).

93. Babenkov, S. V., Aristov, V. Y., Molodtsova, O. V., Winkler, K., Glaser, L., Shevchuk, I., Scholz, F., Seltmann, J., and Viefhaus, J., A new dynamic-XPS end-station for beamline P04 at PETRA III/DESY, *NIM Phys. Res. A*, **777**, 189–193 (2015).

94. Derycke, V., Soukiassian, P., Mayne, A., and Dujardin, G., Scanning tunneling microscopy investigation of the C-terminated β-SiC(100) c(2×2) surface reconstruction: dimer orientation, defects and antiphase boundaries, *Surf. Sci.*, **446**, L101–L107 (2000).

95. Derycke, V., Soukiassian, P., Mayne, A., Dujardin, G., and Gautier, J., Carbon atomic chain formation on the β-SiC(100) surface by controlled sp–sp3 transformation, *Phys. Rev. Lett.*, **81**, 5868–5871 (1998).

96. Semond, F., Soukiassian, P., Mayne, A., Dujardin, G., Douillard, L., and Jaussaud, C., Atomic structure of the β-SiC(100)-(3 x 2) Surface, *Phys. Rev. Lett.*, **77**, 2013–2016 (1997).

97. Aristov, V. Y., Douillard, L., Fauchoux, O., and Soukiassian, P., Temperature-induced semiconducting c (4 x 2)–metallic c (2 x 1) reversible phase transition on the β-SiC(100) surface, *Phys. Rev. Lett.*, **79**, 3700–3703 (1997).

98. Douillard, L., Fauchoux, O., Aristov, V., and Soukiassian, P., Scanning tunneling microscopy evidence of background contamination-induced 2x1 ordering of the β-SiC(100)–c(4 x 2) surface, *Appl. Surf. Sci.*, **166**, 220–223 (2000).

99. Soukiassian, P., Semond, F., Douillard, L., Mayne, A., Dujardin, G., Pizzagalli, L., and Joachim, C., Direct observation of a β-SiC(100)–c(4 × 2) surface reconstruction, *Phys. Rev. Lett.*, **78**, 907–910 (1997).

100. Douillard, L., Aristov, V. Y., Semond, F., and Soukiassian, P., Pairs of Si atomic lines self-assembling on the β-SiC(100) surface: an 8×2 reconstruction, *Surf. Sci.*, **401**, L395–L400 (1998).

101. Hupalo, M., Conrad, E. H., and Tringides, M. C., Growth mechanism for epitaxial graphene on vicinal 6H-SiC(0001) surfaces: a scanning tunneling microscopy study, *Phys. Rev. B*, **80**, 041401 (2009).

102. Wang, Q., Zhang, W., Wang, L., He, K., Ma, X., and Xue, Q., Large-scale uniform bilayer graphene prepared by vacuum graphitization of 6H-SiC(0001) substrates, *J. Phys.: Condens. Matter*, **25**, 095002(4) (2013).

103. Silly, M. G., Roy, J., Enriquez, H., Soukiassian, P., Crotti, C., Fontana, S., and Perfetti, P., Initial oxide/SiC interface formation on C-terminated β-SiC(100) c(2×2) and graphitic C-rich β-SiC(100) 1×1 surfaces, *J. Vac. Sci. Technol. B*, **22**, 2226–2232 (2004).

104. Rosenberg, R. A., Love, P. J., and Rehn, V., Polarization-dependent C(K) near-edge x-ray-absorption fine structure of graphite, *Phys. Rev. B*, **33**, 4034–4037 (1983).

105. Preobrajenski, A. B., Ng, M. L., Vinogradov, A. S., and Mårtensson, N., Controlling graphene corrugation on lattice-mismatched substrates, *Phys. Rev. B*, **78**, 073401 (2008).

106. Meyer, J. C., Geim, A. K., Katsnelson, M. I., Novoselov, K. S., Booth, T. J., and Roth, S., The structure of suspended graphene sheets, *Nature*, **446**, 60–63 (2007).

107. Fasolino, A., Los, J. H., and Katsnelson, M. I., Intrinsic ripples in graphene, *Nat. Mater.*, **6**, 858–861 (2007).

108. Hibino, H., Kageshima, H., Maeda, F., Nagase, M., Kobayashi, Y., and Yamaguchi, H., Microscopic thickness determination of thin graphite films formed on SiC from quantized oscillation in reflectivity of low-energy electrons, *Phys. Rev. B*, **77**, 075413 (2008).

109. Riedl, C., Coletti, C., Iwasaki, T., Zakharov, A. A., and Starke, U., Quasi-free-standing epitaxial graphene on SiC obtained by hydrogen intercalation, *Phys. Rev. Lett.*, **103**, 246804 (2009).

110. Huang, P. Y., Ruiz-Vargas, C. S., van der Zande, A. M., Whitney, W. S., Levendorf, M. P., Kevek, J. W., Garg, S., Alden, J. S., Hustedt, C. J., Zhu, Y., Park, J., McEuen, P. L., and Muller, D. A., Grains and grain boundaries in

single-layer graphene atomic patchwork quilts, *Nature*, **469**, 389–392 (2011).

111. Tao, C., Jiao, L., Yazyev, O. V., Chen, Y.-C., Feng, J., Zhang, X., Capaz, R. B., Tour, J. M., Zettl, A., Louie, S. G., Dai, H., and Crommie, M. F., Spatially resolving edge states of chiral graphene nanoribbons, *Nat. Phys.*, **7**, 616–620 (2011).

112. Tapaszto, L., Dobrik, G., Lambin, P., and Biro, L. P., Tailoring the atomic structure of graphene nanoribbons by scanning tunnelling microscope lithography, *Nat. Nanotechnol.*, **3**, 397–401 (2008).

113. Klimov, N. N., Jung, S., Zhu, S., Li, T., Wright, C. A., Solares, S. D., Newell, D. B., Zhitenev, N. B., and Stroscio, J. A., Electromechanical properties of graphene drumheads, *Science*, **336**, 1557–1561 (2012).

114. Zan, R., Muryn, C., Bangert, U., Mattocks, P., Wincott, P., Vaughan, D., Li, X., Colombo, L., Ruoff, R. S., Hamilton, B., and Novoselov, K. S., Scanning tunnelling microscopy of suspended graphene, *Nanoscale*, **4**, 3065–3068 (2012).

115. Gross, L., Mohn, F., Moll, N., Schuler, B., Criado, A., Guitian, E., Pena, D., Gourdon, A., and Meyer, G., Bond-order discrimination by atomic force microscopy, *Science*, **337**, 1326–1329 (2012).

116. Mashoff, T., Pratzer, M., Geringer, V., Echtermeyer, T. J., Lemme, M. C., Liebmann, M., and Morgenstern, M., Bistability and oscillatory motion of natural nanomembranes appearing within monolayer graphene on silicon dioxide, *Nano Lett.*, **10**, 461–465 (2010).

117. Ondracek, M., Pou, P., Rozsival, V., Gonzalez, S., Jelinek, P., and Perez, R., Forces and currents in carbon nanostructures: are we imaging atoms?, *Phys. Rev. Lett.*, **106**, 176101(4) (2011).

118. Grushko, V. I., Lubben, O., Chaika, A. N., Novikov, N., Mitskevich, E., Chepugov, A., Lysenko, O., Murphy, B. E., Krasnikov, S. A., and Shvets, I. V., Atomically resolved STM imaging with a diamond tip: simulation and experiment, *Nanotechnology*, **25**, 025706(11) (2014).

119. Shirley, E. L., Terminello, L. J., Santoni, A., and Himpsel, F. J., Brillouin-zone-selection effects in graphite photoelectron angular distributions, *Phys. Rev. B*, **51**, 13614–13622 (1995).

120. Menezes, M. G., Capaz, R. B., and Louie, S. G., *Ab initio* quasiparticle band structure of ABA and ABC-stacked graphene trilayers, *Phys. Rev. B*, **89**, 035431 (2014).

121. Marchenko, D., Varykhalov, A., Scholz, M. R., Sánchez-Barriga, J., Rader, O., Rybkina, A., Shikin, A. M., Seyller, Th., and Bihlmayer, G., Spin-

resolved photoemission and *ab initio* theory of graphene/SiC, *Phys. Rev. B*, **88**, 075422(5) (2013).

122. Marchenko, D., Sánchez-Barriga, J., Scholz, M. R., Rader, O., and Varykhalov, A., Spin splitting of Dirac fermions in aligned and rotated graphene on Ir(111), *Phys. Rev. B*, **87**, 115426–8 (2013).

123. Marchenko, D., Varykhalov, A., Scholz, M. R., Bihlmayer, G., Rashba, E. I., Rybkin, A., Shikin, A. M., and Rader, O., Giant rashba splitting in graphene due to hybridization with gold, *Nat. Commun.*, **3**, 1232(6) (2012).

124. Yazev, O. V., and Louie, S. G., Electronic transport in polycrystalline graphene, *Nat. Mater.*, **9**, 806–809 (2010).

125. Banhart, F., Kotakoski, J., and Krasheninnikov, A. V., Structural defects in graphene, *ACS Nano*, **5**, 26–41 (2011).

126. Thrower, P. A., and Mayer, R. M., Point defects and self-diffusion in graphite, *Phys. Status Solidi A*, **47**, 11–37 (1978).

127. Ferry, D. K., and Goodnick, S. M., *Transport in Nanostructures*, Cambridge University Press, Cambridge (1997).

128. Datta, S., *Electronic Transport in Mesoscopic Systems*, Cambridge University Press, Cambridge (1995).

Chapter 3

Graphene Growth via Thermal Decomposition on Cubic SiC(111)/Si(111)

B. Gupta,[a] N. Motta,[a] and A. Ouerghi[b]

[a]*School of Chemistry, Physics and Mechanical Engineering and Institute for Future Environments, Queensland University of Technology, 2 George Street, Brisbane 4001, QLD, Australia*
[b]*Laboratory for Photonics and Nanostructures - LPN- CNRS, Route de Nosay, Marcoussis 91460, France*
n.motta@qut.edu.au

3.1 Introduction

Graphene has been often represented as the "holy grail" of electronic miniaturization [1, 2], the new material to replace Si in order to extend the Moore law beyond the present limit [3] and achieve increased performance and energy saving. Some of the expected applications of graphene include high-frequency transistors, reversible hydrogen storage, lithium ion batteries [4], supercapacitors [5], sensors [6, 7], biotechnology [8], and transparent electrode in solar cells [9],

---

Growing Graphene on Semiconductors
Edited by Nunzio Motta, Francesca Iacopi, and Camilla Coletti
Copyright © 2017 Pan Stanford Publishing Pte. Ltd.
ISBN 978-981-4774-21-5 (Hardcover), 978-1-315-18615-3 (eBook)
www.panstanford.com

to mention just a few [2]. However, the absence of a bandgap limits severely its application in high-performance integrated logic circuits as a planar channel material. Several routes have been followed to induce and control such a gap in graphene by using chemical doping, quantum dots, nanoribbons, and functionalization [10–14]. An interesting possibility is to exploit the opening of a gap caused by the interaction of graphene with a SiC substrate [15], making the epitaxial growth of graphene on SiC by Si sublimation a very attractive strategy to achieve this goal. SiC is also compatible with the Si technology, and the direct growth of graphene on SiC would open the way for the direct integration of graphene in large-scale electronics. As SiC is expensive, and to improve the compatibility with the Si industry, the growth of graphene on a thin layer of SiC grown on Si has been proposed [16].

In this chapter we discuss the growth of epitaxial graphene on 3C-SiC(111)/Si(111) and the results obtained by several authors in this research area. Due to the hexagonal structure formed by the cubic (111) face of cubic 3C-SiC, graphene is expected to grow epitaxially. However, the large lattice parameter difference should be considered in determining the growth model. In the following we will review the details of transformation of SiC to graphene at the atomic level, how graphene growth proceeds with time, and how to produce continuous large areas of graphene on 3C-SiC(111)/Si(111).

3.2 Epitaxial Growth of Graphene

The term "epitaxy" comes from Greek, meaning the growth or deposition of a crystalline overlayer on a crystalline substrate. The overlayer is considered as the epitaxial layer. In this section we will discuss first the growth of epitaxial graphene layer on a bulk SiC substrate.

3.2.1 Thermal Graphitization of the SiC Surface

The formation of graphite by ultrahigh vacuum (UHV) annealing of the SiC surface had been studied many years before the discovery of graphene by Badami in 1961 [17] and, more recently, by Forbeaux et al. in 2000 [18]. However, none of these experiments provided evidence of 2D crystals.

The thermal graphitization of bulk SiC involves heating the sample to temperatures between 1200°C and 1600°C, resulting in the thermal decomposition of the surface [19]. At these temperatures, the silicon sublimates from the surface, while the remaining carbon atoms undergo diffusion and nucleation to form the epitaxial graphene layers. Pioneering work on the production of graphene from bulk 6H and 4H SiC has been done by Berger, Sutter, Emtsev, de Heer, and Ouerghi [20–25]. The annealing of bulk SiC (6H/4H) can be performed in an argon (Ar) atmosphere or in a UHV chamber. In an Ar atmosphere, the temperature for graphene growth is quite high in comparison to the growth in UHV. The Ar pressure on the surface produces a decrease in the Si sublimation rate, and therefore a controlled number of graphene layers is easily obtained. In UHV the growth rate is quite high, as without a counteracting pressure it is easy for Si to break the bonds and sublimate, resulting often in an uncontrolled growth of a large number of graphene layers.

Emtsev et al. compared these two growth processes [24]. Before the final annealing to obtain graphene they used hydrogen etching on the substrates in order to remove any kind of polishing damage from the surface. This method produces a highly uniform, flat surface with wide steps, which is beneficial for graphene growth. They showed that the graphene layer grown on bulk SiC in an Ar atmosphere is composed of larger crystals and the film quality is improved compared to the graphene grown in UHV. De Heer et al. suggested a method called confinement-controlled sublimation (CCS), by enclosing the SiC in a small quartz vessel [26]. They used as well hydrogen etching to remove the polishing damage before the growth and compared the growth process and quality of graphene from the UHV and CCS methods. They concluded that the graphene layers produced by the CCS method are defect free, and they were able to produce a uniform monolayer (ML) on both polar faces. The key of the CCS method is the control of the silicon vapor density, which ensures a near-thermodynamic equilibrium, essential for a uniform growth. In addition, it allows good control of the graphitization temperature, which is important to control the graphene growth. Later on, Ouerghi et al. [25] found that off-axis 6H SiC(0001) can provide high-quality graphene over a large area, by exploiting the step bunching effect. Graphene starts growing on the steps and slowly grows on the terraces. The step bunching slows down the formation of a continuous graphene layer, but it provides a way to control the number of layers. They used 6H SiC(0001) 3.5° off toward (11-20) and exposed the sample to a N_2 partial pressure of $P = 2 \times 10^{-5}$ torr and a Si deposition rate of 1 mL/min during the graphene growth, trying to reproduce conditions similar to the graphene growth in Ar. Because of N_2 pressure and Si deposition, Si atoms from the SiC substrate could not sublimate quickly, as would be expected in UHV, producing uniform mono- or bilayer graphene.

3.2.2 Graphene Growth on Cubic SiC(111)/Si(111)

Graphene produced on bulk SiC annealing presents interesting electrical properties [27, 28], but it is expensive as compared to the standard Si substrate. For this reason, large industrial-scale fabrication and mass production of graphene will not be possible using bulk SiC.

To address this issue, heteroepitaxial growth of cubic polytype 3C-SiC on silicon substrates has been proposed [16]. High-quality 3C-SiC/Si is challenging to achieve in terms of continuity, film thickness uniformity, and defect density arising from lattice/ thermal mismatch between Si and SiC [29]. Large efforts have been made to improve these aspects, and a smooth and uniform 3C-SiC/ Si surface over large areas has been achieved recently [29, 30]. The proposed orientations of 3C-SiC on Si for the synthesis of epitaxial graphene are (111), (100), (110), and (001) [31–38]. Among these orientations, (111) is preferred over (110) and (100). The surface atomic arrangement of 3C-SiC(100) is made of square centered units but is affected by antiphase domains (APDs). So the growth of graphene on this type of substrate results in two- or multicrystalline orientations [39]. In spite of this, recently good-quality graphene has been obtained on off-axis 3C-SiC(100) [39]. One important fact related to 3C-SiC(100) substrates is that this polytype does not form terraces as the 3C-SiC(111)/6H/4H SiC. This might be an advantage for the continuity of graphene layers, but a high number of defects could be present due to the APD.

The advantage of growing 3C-SiC(111) on Si(111) is threefold. First of all, the Si substrate is cheaper and available in substantially larger areas than the commercial SiC wafers, fully compatible with the current Si processing techniques, making graphene technology more attractive from an industrial point of view. Second, the graphitization of 3C-SiC epilayers on Si produces ML, bilayer, or multilayer graphene [21, 40] on a Si platform, which can be easily integrated in the electronic industry without the need of complex transfers or chemical etching steps [24]. Third, the graphitization on the surface of 3C-SiC(111) proceeds in a similar manner to that on the hexagonal SiC bulk crystals [35, 41], with two polar faces like the 6H and 4H [42, 43]. The epitaxial growth of graphene on the Si-terminated face of 3C-SiC(111) should be similar to that on Si-terminated 6H SiC(0001) [43, 44] as the top four atomic layers of both surfaces are identical (Fig. 3.1) [35]. This is true for ML graphene. However, for multilayer graphene it has been found that the cubic SiC favors the ABC stacking, while the hexagonal SiC favors the ABA stacking, with different electronic properties [45, 46].

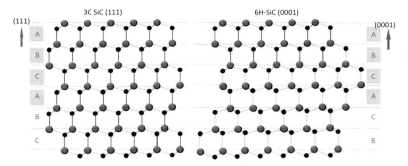

Figure 3.1 Crystallographic surface structure of 3C-SiC(111) (left) and 6H SIC(0001) (right) [47].

epitaxial graphene growth on the 3C-SiC(111)/Si(111) substrates has been achieved by Suemitsu et al. [36], Ouerghi et al. [43], and Gupta et al. [48]. The study of graphene on Si-terminated 3C-SiC(111)/Si(111) was started by Ouerghi [31, 40, 43] in 2010 by using UHV high-temperature annealing; he followed and explained the different reconstructions that occur at increasing temperatures [31, 40]. He studied the growth, structure, and electronic properties of the resulting graphene via scanning tunneling microscopy (STM), X-ray photoelectron spectroscopy (XPS), Raman spectroscopy, angle-resolved photoemission spectroscopy (ARPES), and low-energy electron diffraction (LEED) on a 600 nm thick 3C-SiC(111) layer grown on Si(111) substrates in order to prevent stress-induced cracks at the surface. Later, this work was extended by Gupta et al. on a 250 nm thick 3C-SiC(111) layer [47, 48].

To produce graphene the SiC/Si substrates were first degassed for several hours at 600°C under UHV conditions and then annealed under a low (0.1 nm/min) Si flux at 900°C to remove the native oxide. This step was followed by annealing of the substrates at temperatures ranging from 900°C to 1300°C.

3.3 Surface Transformation: From 3C-SiC(111)/ Si(111) to Graphene

While the different reconstructions that appear on the SiC surface at different steps of the annealing process are well known, the details of the transformation from SiC to graphene are not completely clear. As the 3C-SiC(111) epilayer is several hundreds of nanometers thick, the surface reconstructions occurring during annealing are substantially unaffected by the presence of the Si substrate. So we will start the discussion from the process of reconstruction occurring on bulk SiC according to the literature.

3.3.1 Reconstructions of SiC(111)

The graphitic structures obtained by high-temperature annealing of SiC and its different stages of reconstruction have been studied by LEED since 1975 [49]. It has been recognized that the reconstructions leading to graphene are different for silicon (Si)-terminated and carbon (C)-terminated faces of 6H and 4H SiC. For both the Si- and C-terminated face, the reconstruction proceeds by increasing the annealing temperature and time.

3.3.1.1 Si-terminated face

In this case it goes through (3×3), $(\sqrt{3} \times \sqrt{3})R30°$, $(6\sqrt{3} \times 6\sqrt{3})R30°$, and graphene (1×1).

The first two reconstructions (Fig. 3.2a,b) on SiC are strongly dependent on the amount of Si supply, the heating time, and the heating temperature [52–54]. (3 × 3) is a Si-rich phase and has a very stable structure that consists of a complete Si adlayer on top of an uppermost bulk-like SiC substrate layer, which is usually obtained by annealing at 850°C under Si flux. The adlayer contains no vacancies or corner holes, and it is covered by tetrahedral adatom clusters with three Si base atoms and one top Si atom per unit cell [50]; the base Si trimer lies on a twisted Si adlayer, forming cloverlike rings on the Si-terminated face. On the other hand, $(\sqrt{3} \times \sqrt{3})R30°$ reconstruction is favored by less Si-rich preparation conditions [55]. The $(6\sqrt{3} \times 6\sqrt{3})R30°$ phase has a complicated surface reconstruction and consists of (13 × 13) unit cells of graphene (Fig. 3.2c) . This surface reconstruction is attributed to a C-rich phase but does not have any graphitic properties as the adlayer has a strong interaction with the substrate and so it is considered a buffer layer or an interface layer. The buffer layer passivates the SiC surface so that the subsequent C planes are only slightly attached to the substrate [51, 56, 57]. Finally, the (1 × 1) graphene surface structure appears when the formation of the graphene layer is complete on the surface. The above surface reconstructions do not depend on the particular SiC polytype, and they have been been found on all the hexagonally arranged surfaces of 3C, 4H, and 6H SiC [22, 43, 56, 58–61]. Interestingly, the $(6\sqrt{3} \times 6\sqrt{3})R30°$ is not observed on the C-terminated face, while the $(\sqrt{3} \times \sqrt{3})R30°$ has been rarely reported [62].

3.3.1.2 C-terminated face

The most accepted sequence of graphene growth on the C-terminated face is $(2 \times 2)_{Si}$, (3 × 3), $(2 \times 2)_C$ and (1 × 1) graphene [56, 57, 63–66]. The two phases $(2 \times 2)_{Si}$ and $(2 \times 2)_C$ occur from different surface treatments, leading to Si- and C-rich structures, respectively [56, 66]. The $(2 \times 2)_{Si}$ phase develops upon annealing in Si flux. This procedure also helps to remove surface oxides from the SiC samples [48, 56]. Further annealing of the surface leads toward (3 × 3) and $(2 \times 2)_C$ surface reconstructions [56, 63] and finally to graphene growth.

The best terminated face of SiC for graphene growth is still debatable as both faces have pros and cons, including the electronic properties of graphene. It is different for both faces. On the Si-terminated face, an interface layer with sp^3 configuration is present and the atoms in this plane form strong covalent bonds with the SiC substrate. Therefore, graphene is strongly bonded to the substrate due to the charge transfer from the interface layer (buffer layer) to the graphene layers, which results in a doping effect [62, 67]. It has been found by experimental and theoretical work [64, 66] that graphene on a (3 × 3) structure is less bound to the substrate in comparison to the ($6\sqrt{3}$ × $6\sqrt{3}$) structure. Moreover, it has been shown that the C-terminated face of the polar SiC(000-1) supports graphene growth without altering its electronic properties or with a negligible amount of interaction in comparison with the Si-terminated face [21, 68]. In addition, on the C face epitaxial graphene grows with a certain rotational disorder where adjacent layers are rotated relative to each other. Stacking faults decouple adjacent graphene sheets so that their band structure is nearly identical to the isolated graphene. In fact the Dirac dispersion at the K-point is preserved even though the film is composed of many graphene sheets [27]. In the case of the SiC(0001)/Si surface, the graphitization process is slow and the number of graphene layers (usually one or two) can be controlled more easily. Unfortunately, the resulting electron mobility turns out to be rather low, so the Si-terminated surface is not suited to the fabrication of samples used in transport measurements. For the C-terminated (000-1) surface the graphitization process is very fast and a large number of graphene layers can be formed quickly (up to 100). However, in this case the electron mobility is rather high [26].

A number of studies reported about the growth of epitaxial graphene on 3C-SiC [31, 32, 35, 40, 43, 48, 69–71], but so far the details of the atomic transformation leading to the formation of graphene have not yet been clarified. 3C-SiC(111)/Si(111) also has two polar faces (6H and 4H). Little attention has been given to the C face, while the Si face has been studied comparably well. Si-terminated 3C-SiC/Si(111) follows the same sequence of reconstuctions with temperature toward graphene as bulk 6H SiC(0001) or 4H SiC(0001) [36, 43, 63, 71]. In the following we will analyze the reconstructions and formation of epitaxial graphene layers on 3C-SiC(111) through LEED, low-energy electron microscopy (LEEM), and STM studies.

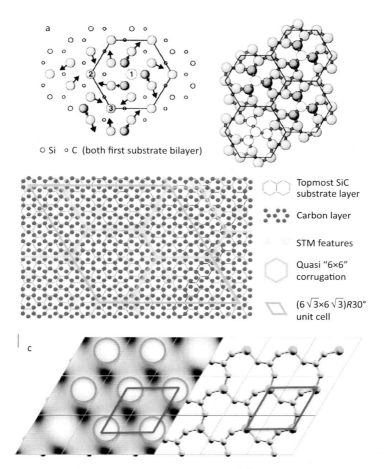

Figure 3.2 Surface reconstructions of SiC at different stages of graphene formation (a) 3 × 3, (b) ($\sqrt{3} \times \sqrt{3}$)R30°, and (c) ($6\sqrt{3} \times 6\sqrt{3}$)R30°. (a) Reprinted (figure) with permission from Ref. [50]. Copyright (1998) by the American Physical Society. (b) From Ref. [51]. @ IOP Publishing. Reproduced with permission. All rights reserved. (c) From Ref. [47].

3.3.2 LEED and LEEM Characterization of the Transformation

The growth of epitaxial graphene on a thin film of SiC(111) on Si(111) was obtained by a few groups [36, 43, 48] on unpolished 3C-SiC(111) epilayers on Si(111) substrates, with typical roughness of 0.5 nm. The epitaxial graphene was prepared in UHV at a pressure of 2×10^{-10} mbar. Figure 3.3 shows the LEED patterns of surface reconstructions related to different temperatures obtained by Ouerghi et al. [31]. Annealing at 800°C for 15 min under Si flux resulted in the formation of a (3 × 3) surface reconstruction, the Si-rich phase (Fig. 3.3a). After 900°C annealing for 5 minm a formation of a (6 × 6) reconstructed surface was obtained and assigned to a Si-rich surface (Fig. 3.3b). A $(\sqrt{3} \times \sqrt{3})R30°$ (only a 1/3 ML Si atom remains on the surface) super structure was obtained by annealing at 1050°C (Fig. 3.3c). A mixture of $(\sqrt{3} \times \sqrt{3})R30°$ and $(6\sqrt{3} \times 6\sqrt{3})R30°$ surface was produced upon annealing at 1100°C (Fig. 3.3d). At 1200°C, the formation of a complete $(6\sqrt{3} \times 6\sqrt{3})R30°$ super structure was observed (Fig. 3.3e). Graphene (C termination) and a few layers of graphene were obtained by annealing the sample at 1250°C and 1300°C for 10 min (Fig. 3.3f).

It must be mentioned that the C-terminated face of 3C-SiC(111) has not been studied as widely as the Si face of 3C-SiC(111). The C-terminated face of 3C-SiC(111)/Si(111) had not been considered until Fukidome et al. [72], who studied the epitaxial graphene growth on both Si and C faces of 3C-SiC(111). They showed how the surface termination of 3C-SiC(111)/Si affects the stacking, the interface structure, and the electronic properties of graphene. They used 3C-SiC(111)/Si(111) to study the growth on the Si face and 3C-SiC(111)/Si(110) (rotated epitaxy) to study the growth on the C face. The epitaxial graphene was grown on both surfaces at 1250°C for 30 min in vacuum. For the 3C-SiC(111)/Si(111) surface, LEED hexagonal spots were clearly observed, indicating the presence of a Bernal stacked epitaxial graphene on 3C-SiC(111), similar to epitaxial graphene on 6H SiC(0001). For the rotated 3C-SiC(111)/Si(110) thin film, on the other hand, the LEED spots were smeared out and formed modulated diffraction rings (Fig. 3.4a,b). This indicates turbostratic stacking of epitaxial graphene in the same manner as epitaxial graphene on 6H SiC(000-1) [63, 73]. The variation of the stacking could be related to the change in the interface structure, that is, the presence/absence of the buffer layer that works as a template for the epitaxy of graphene.

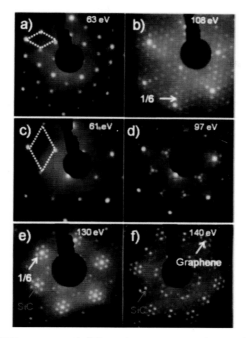

Figure 3.3 LEED patterns of differently reconstructed 3C-SiC(111) obtained by increasing the annealing temperature without Si flux (a) (3 × 3) phase after annealing at 800°C, (b) (6 × 6) phase after annealing at 900°C, (c) ($\sqrt{3}$ × $\sqrt{3}$) R30° phase after annealing at 1050°C, (d) ($\sqrt{3}$ × $\sqrt{3}$)R30°(6$\sqrt{3}$ × 6$\sqrt{3}$)R30° after annealing at 1100°C, (e) (6$\sqrt{3}$ × 6$\sqrt{3}$)R30° after annealing at 1200°C, and (f) epitaxial graphene layer after annealing at 1250°C. Reprinted (figure) with permission from Ref. [31]. Copyright (2010) by the American Physical Society.

(a)   (b)

Figure 3.4 Characterization of epitaxial graphene on nonrotated 3C-SiC(111)/Si(111) and rotated 3C-SiC(111)/Si(110). (a) LEED pattern of the epitaxial graphene on 3C-SiC(111)/Si(111); the incident electron energy is 58 eV. (b) LEED pattern of the epitaxial graphene on 3C-SiC(111)/Si(110) using the electron optics of LEEM; the incident energy is 55 eV. Reproduced with permission from Ref. [72]. Copyright (2011) The Japan Society of Applied Physics.

Darakchieva et al. also studied the graphene growth on both polarities of 3C-SiC(111) [42]. A few-hundred microns of 3C-SiC layers were grown on 6H SiC(0001) by sublimation epitaxy. Epitaxial graphene layers were grown on the Si face and C face of 3C-SiC(111) via high-temperature sublimation in an Ar atmosphere under optimized conditions (the exact temperature, time, and the graphene growth details are not explained). Figure 3.5 shows LEEM images and the LEED pattern of these polar faces. The Si face (Fig. 3.5a) shows uniform coverage of single-layer graphene. The presence of bilayer graphene is very limited. On the other hand the C face is covered by nonuniform graphene layers (Fig. 3.5b).

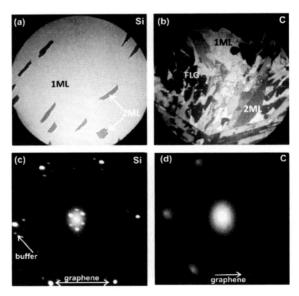

Figure 3.5 LEEM images from selected sample areas for EG on (a) the Si face and (b) the C face 3C-SiC (field of view 50 µm). Domains with 1-, 2-, and few-monolayer (FML: 3 and 4 monolayers) graphene are indicated on the LEEM images. µ-LEED pattern from a single-layer graphene (ML) area of the (c) Si face and (d) C face taken at 40 and 44 eV, respectively. Reprinted from Ref. [42], with the permission of AIP Publishing.

LEED patterns (Fig. 3.5c) revealed 1 × 1 diffraction spots associated with the single-layer graphene surrounded by the $(6\sqrt{3} \times 6\sqrt{3})R30°$ diffraction spots connected with the SiC surface. On the C-terminated face of 3C-SiC(111) (Fig. 3.5d), only the graphene diffraction spots are visible. This indicates a different interface structure compared to the one present on the Si face. Their studies indicated that the predominant type of defect on the C face was different from the twin boundaries found on the Si face. Small inclusions occurred on the C face of 3C-SiC(111), each associated with 6H SiC formed around a screw dislocation. In summary they achieved a homogeneous growth of graphene on an approximately 2 × 2 mm^2 area of the Si face. These domains are considerably smaller on the C face.

3.3.3 STM Characterization: Atomic Resolution Imaging of the Transition

Gupta et al. analyzed recently by STM the transformation of 3C-SiC(111) into graphene due to high-temperature annealing in UHV [47]. They provided a clear picture of the sequence of reconstructions leading to graphene formation on 3C-SiC(111) with the help of STM and density functional theory–local density approximation (DFT-LDA) calculations. Thermal graphitization was achieved by annealing the substrate at 1250°C in UHV ($p < 5\cdot10^{-10}$ mbar) for 10′.

Figure 3.6a shows the coexistence of two different reconstructions, with a progressive transition from one to the other. The transiton evolves from the right to the left of the image. The brighter atoms are marked with blue circles in the left region and with black circles in the right region (Fig. 3.6b). Hollow hexagons with three brighter and three darker atoms are visible, with a lattice constant of 5.3 ± 0.05 Å (Fig. 3.6c), very close to $\sqrt{3}a = 5.35$ Å, which indicates a($\sqrt{3}$ × $\sqrt{3}$)$R30°$ surface reconstruction of the SiC. On the left side of the image the central holes and the hexagons have shrunk and the atoms now appear arranged in parallel lines with alternated dark and bright atoms. The periodicity along and across these lines is 5.2 ± 0.05 Å and 4.6 ± 0.05 Å, respectively, matching a nominal $\left(\dfrac{3}{2}\times\sqrt{3}\right)$ $R30°$ surface reconstruction (Fig. 3.6d).

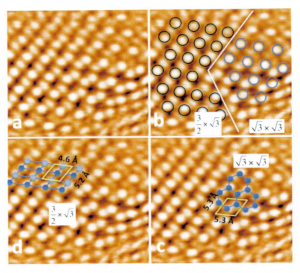

Figure 3.6 (a) (3.6 × 3.6)nm² STM image (V = 60 mV; I = 80 pA) of **3C-SiC(111)** after annealing at 1250°C, showing the coexistence of two different reconstructed phases. (b) The same image with black and blue circles around the brighter atoms to indicate the two different reconstructions. (c) The ($\sqrt{3}$ × $\sqrt{3}$)$R30°$ reconstruction. (d) The $\left(\frac{3}{2} \times \sqrt{3}\right) R30°$ reconstruction. The panels are labeled following the direction of reconstruction [76].

To validate the model the authors simulated the STM images by using DFT-LDA calculations [74]. The model converged for the ($\sqrt{3}$ × $\sqrt{3}$) reconstruction but did not detect any stable minimum of the energy for the $\left(\frac{3}{2} \times \sqrt{3}\right) R30°$. The authors suggest that this latter reconstruction is incommensurate with the SiC periodicity and thereby impossible to simulate within periodic boundary conditions. This would also entail a high formation energy and a lack of stability for this structure, which is only a transition toward graphene [75].

The transformation of $\left(\frac{3}{2} \times \sqrt{3}\right) R30°$ to graphene is presented in Fig. 3.7 [47]. A residual of the $\left(\frac{3}{2} \times \sqrt{3}\right)$ reconstruction is evidenced by the left yellow rhombus in Fig. 3.7c. The bright atoms' distances in the rhombus are now 4.9 ± 0.05 Å and 4.6 ± 0.05 Å, with

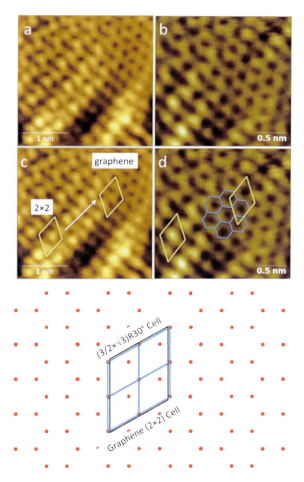

Figure 3.7 Transformation of the $\left(\frac{3}{2} \times \sqrt{3}\right) R30°$ unit cell into a 2 × 2 cell and to graphene. (a) STM image (bias, 0.02 V; current, 80 pA) of the (2 × 2) region—former $\left(\frac{3}{2} \times \sqrt{3}\right) R30°$—where it is possible to see the transformation going from bottom left to top right. (b) STM close-up image of the graphene region (bias, 0.02 V; current, 80 pA). (c) Image (a) with the 2 × 2 structures superimposed. (d) Image (b) where the blue hexagons with a 2.45 ± 0.05 Å periodicity indicate the graphene area. (e) A comparision between $\left(\frac{3}{2} \times \sqrt{3}\right) R30°$ unit cell (black) and a graphene (2 × 2) cell (blue) [76].

5.2 Å compressed to 4.9 Å. The latter distance corresponds to the periodicity of the graphene (2 × 2) reconstruction, strengthening the hypothesis that the $\left(\frac{3}{2} \times \sqrt{3}\right)$ is a precursor of the graphene formation. Figure 3.7e shows a tentative overlap of the $(4.6 \times 5.2)\text{Å}^2$ cell with the graphene (2 × 2) lattice cell, proving that the distortion required to transform the $\left(\frac{3}{2} \times \sqrt{3}\right)$ to a (2 × 2) is very small, as the surface ratio between the two structures is 0.99. The ratio of $(\sqrt{3} \times \sqrt{3})R30°$ to the graphene lattice is instead 1.18, involving a greater distortion; as a consequence, the direct transition from $(\sqrt{3} \times \sqrt{3})R30°$ to graphene cannot occur and intermediate reconstruction $\left(\frac{3}{2} \times \sqrt{3}\right)$ is required.

It is interesting to observe the gradual appearance of the graphene structure going from the bottom left to the top right of the image (Fig. 3.7c), as more atoms leave the reconstruction. At the top-right end (the center in Fig. 3.7b) it is possible to see clearly the (1 × 1) cell of the ML graphene with a periodicity of 2.45 ± 0.05 Å (marked in Fig. 3.7d).

As the stacking sequence of the first four atomic planes in 3C-SiC(111) is similar to that of 6H and 4H SiC(0001) it is suggested that this model is applicable to any transformation of SiC into graphene [47].

3.3.4 Atomic Structure Studies of Bi- and Multilayer Graphene

The atomic structure of thicker graphene layers epitaxially grown on 3C-SiC(111)/Si(111) in UHV [48] has been also investigated. STM images obtained after annealing the substrate at 1300°C are shown in Fig. 3.8. The graphene layer appears to be continuous on the substrate, with the presence of wrinkles because of the steps and defects in the underlying SiC/Si(111). The STM image is showing a step where a moiré pattern is visible with a periodicity of 17 Å. According to literature, this structure is attributed to the C-rich $(6\sqrt{3} \times 6\sqrt{3})R30°$ reconstruction due to the 30° rotation of the graphene overlayer, with respect to the unreconstructed 3C-SiC(111) surface [43, 51, 76]. Figure 3.8b,d shows the typical structure of the Bernal stacking, with a periodicity of 2.45 Å, confirming the presence of more than one graphene layer. The STM image of multilayer graphene at the atomic scale exhibits the typical three-for-six symmetry of graphite.

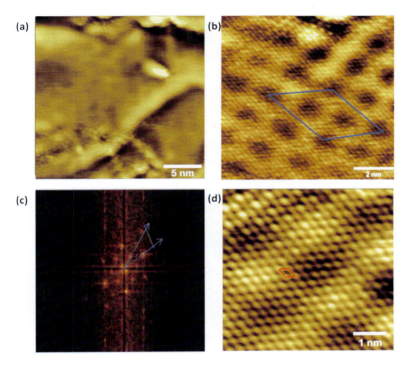

Figure 3.8 STM images of graphene obtained by annealing SiC/Si(111) at 1300°C. (a) 20 × 20 nm^2 area with a step showing a shadow of a moiré pattern (V = 70 mV; I = 0.3 nA). (b) High-resolution moiré pattern with hexagonal symmetry (V = 50 mV; I = 0.2 nA); a $(6\sqrt{3} \times 6\sqrt{3})R30°$ unit cell is also shown. (c) FFT of image (b) showing 27° rotation of the graphene layer with respect to the buffer layer. (d) High-resolution STM image of bi-/few-layer graphene (V = 50 mV; I = 0.2 nA) with a graphene unit cell (red insert) [48].

In the fast Fourier transform (FFT) image (Fig. 3.8c), three sets of bright spots are visible, corresponding to the first- and second-nearest neighbors and to the moiré pattern. From the angle measurements (Fig. 3.8c), it is possible to measure the angle of the buffer layer rotation with respect to the SiC substrate (27° ± 3°). A back Fourier transform of Fig. 3.8b, obtained by selecting the different hexagons of bright spots in Fig. 3.8c, shows the different contributions to the STM image (Fig. 3.9). The innermost, middle, and outermost hexagons in Fig. 3.8c are the Fourier components of the moiré structure, second- and first-nearest neighbor of the three-for-six symmetry of graphite, respectively. The periodicity of these structures as measured in Fig. 3.9a–c gives a value of the spacing equal to 17 Å, 4.2 Å, and 2.46 Å, respectively.

The epitaxial graphene domains are small (10–15 nm), but other authors were able to achieve domains of μm size by using off-axis substrates [39], by hydrogen etching and by using polished substrates (see Section 3.5) [44]. The combination of these methods is the key to reducing the roughness and defects of the substrate. As a result, the sublimation rate will be the same over the whole surface, allowing the production of large-area epitaxial graphene.

Figure 3.9 FFT images extracted by selecting three hexagons of spots in Fig. 3.11c, as shown in the insets: (a) moiré pattern from the innermost hexagon, (b) second-nearest neighbors from the middle hexagon, and (c) first-nearest neighbors (in the three-for-six symmetry of graphene atoms) from the outer hexagon [48].

The larger domains of epitaxial graphene on 3C-SiC(111)/Si(111) were shown by Ouerghi et al. [31] Fig. 3.10, where large flat terraces are present, indicating the presence of continuous graphene layers.

They have also shown the STM and scanning transmission electron microscopy (STEM) images of continuous ML graphene over the step edges (Fig. 3.11a,b), which proves the presence of large area ML graphene on 3C-SiC(111). They found this behavior at many single-atomic-height steps but have not detected any double-atomic-height step of the graphene layer. For this reason they proposed a schematic of an atomic structure graphene layer (Fig. 3.11c). They suggest that bond breaking may occur more easily at step edges, caused by Si-C diffusion or Si desorption, giving origin to graphene layers bypassing the steps. This ensures the continuity of the top layer from ML to bilayer graphene.

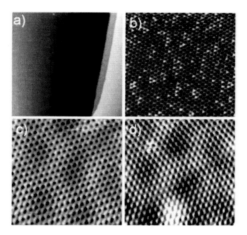

Figure 3.10 STM micrographs of graphene/3C-SiC(111) epilayers; (a) STM images (150 × 150) nm² (−2 V; 0.2 nA), (b) STM images (50 × 50) nm² (−2 V; 0.2 nA), (c) honeycomb-type structures (5 × 5) nm² (−45 mV; 0.2 nA), and (d) triangular-type structures (5 × 5) nm² (−25 mV; 0.2 nA) of the surface of graphene/3C-SiC(111). Reprinted (figure) with permission from Ref. [31]. Copyright (2010) by the American Physical Society.

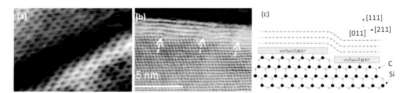

Figure 3.11 (a) A (5 × 5) nm² images of the honeycomb structure of graphene (−0.05 V; 0.1 nA), (b) a high-resolution STEM image of three-layer graphene, and (c) a schematic of the atomic structure of a STEM image showing three-layer graphene on the steps of 3C-SiC(111). Reprinted from Ref. [40], with the permission of AIP Publishing.

3.3.5 Improving the Epitaxial Graphene Quality by Using Polished Substrates

Large-area and defect-free graphene is required for application in electronics and photonics. To this purpose polishing the substrate prior to graphene growth is mandatory. Yazdi et al. obtained graphene growth on the Si-terminated face of a 4H, 6H, and 3C-SiC substrate by Si sublimation from SiC in an Ar atmosphere at a temperature of 2000°C [44]. The monocrystalline Si-terminated 3C-SiC(111) (~1.5 μm) substrates were grown on an on-axis 6H SiC(0001) wafer. It was found that the 3C-SiC(111) polytype produces the best-quality ML graphene. A large area, over 50×50 μm^2, ML graphene was obtained on 3C-SiC(111). Additionally, they compared the results of polished and unpolished surface of 3C-SiC(111) and showed how the roughness of the substrate affects the graphene growth.

Figures 3.12a and 3.12b show LEEM images of graphene grown on unpolished and polished 3C-SiC(111) substrates, respectively. Of the surface, 65% was covered by bilayer graphene on the unpolished substrate and the rest by three-layer graphene. On the polished substrate most of the surface (93%) was covered by ML graphene, with bilayer graphene in very few areas. Roughness measurements taken by AFM (Fig. 3.12c,d) reported 2 nm for the unpolished substrate and 0.6 nm for the polished one. The histogram in Fig. 3.12e shows that the lower roughness on the polished substrate results in less pronounced step bunching and consequently in a better quality of the grown graphene. So the surface roughness should always be minimized before graphene growth. ARPES of the π-band for the polished sample demonstrates a perfect linear dependence, characteristic of 1 ML graphene (the inset in Fig. 3.12f). The sublimation rate of 3C-SiC is the same over the whole defect-free substrate surface, as compared to other polytypes, resulting in a superior uniformity of the grown graphene layer. The authors found that the presence of single Si-C bilayer steps at the beginning of the graphene formation is the controlling factor for a uniform sublimation of Si.

The effect of polishing has been studied on 3C-SiC(111)/Si(111) by Gupta et al. [77]. ARPES of the epitaxial graphene on a polished substrate obtained at 80 K (Fig. 3.13) shows the typical band structure of graphene is obtained, where π-and σ-bands are visible.

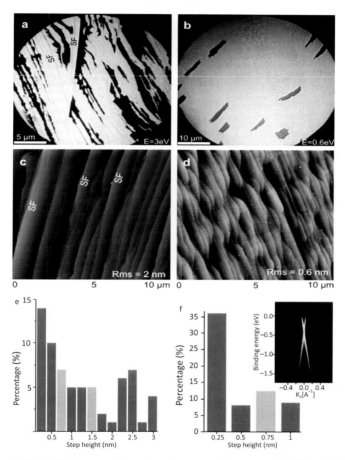

Figure 3.12 LEEM images of graphene layers on (a) as-grown 3C-SiC with ~65% coverage by 2 ML graphene (bright area); the rest is 3 ML. Some of stacking faults (SFs) are shown. (b) Polished 3C-SiC with ~93% coverage by ML graphene (bright area) and ~7% of 2 MLs. AFM images of graphene on (c) as-grown 3C-SiC; some SFs are shown. (d) A polished 3C-SiC substrate. Histograms of step heights for (e) graphene on an as-grown substrate showing a wide distribution of step heights. (f) Graphene on polished substrate. Inset image shows an ARPES spectrum of the π-band taken at the K-point. Reprinted from Ref. [44], Copyright (2013), with permission from Elsevier.

The energy distribution curves (EDCs) relative to the Dirac cone apex are reported in Fig. 3.14a. The π-band peak of each curve was fitted with a Gaussian function, and the center of each Gaussian is indicated by a red star in Fig. 3.14a. The curve connecting the position of the Gaussian maxima as a function of the parallel momentum was, in turn, fitted with an hyperbole, which is the result of the intersection of a cone with a plane not containing a diameter of the cone, and therefore it is the curve we expect to see in the place of the Dirac cone. By fixing the known tilt angle with respect to the $\overline{\Gamma}-\overline{K}$ direction (1.7°, obtained by LEED measurements) it was possible to extrapolate the exact position of the vertex of the Dirac cone E_D. We find it to be located 0.29 eV below the Fermi level. It is worth noting that the shape and width of the bands are very sensitive to the number of graphene layers, as clearly reported in recent ARPES [78] and spatial-resolved nano-ARPES [79] of multilayer graphene on bulk SiC. Thus, the linearity and width of the Dirac cone of the ARPES data in Fig. 3.14c suggest the coexistence of grains of few-layer graphene with different thicknesses and without rotational disorder.

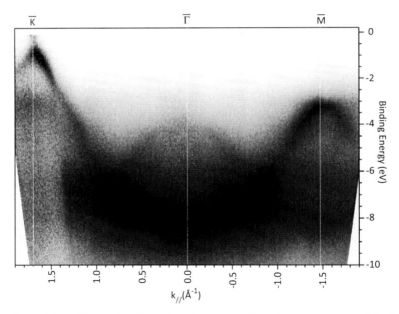

Figure 3.13 Electronic valence-band structure of graphene grown on polished 1 μm 3C-SiC(111)/Si(111) acquired by ARPES at 77 K. Reproduced from Ref. [77]. © IOP Publishing. Reproduced with permission. All rights reserved.

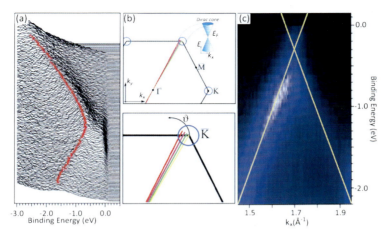

Figure 3.14 (a) Energy distribution curves of the valence band around the apex of the Dirac cone. The red star on each curve indicates the position of the center of the Gaussian peak used to fit the peak of the π-band. (b) Sketch of the graphene Brillouin zone. Lines of different colors correspond to cuts along different azimuthal directions. (c) Close-up of the Dirac cone showing a small shift due to the rotation with respect to the $\bar{\Gamma}-\bar{K}$ direction of our sample. The yellow line is a guide for the eye to locate the Dirac cone. Reproduced from Ref. [77]. © IOP Publishing. Reproduced with permission. All rights reserved.

3.4 Conclusion

This study is a summary of the knowledge acquired in the recent years regarding graphene growth by Si sublimation on 3C-SiC(111)/Si(111) in UHV. This substrate is particularly important as it opens the way to a graphene growth compatible with industrial microelectronic processing. We presented the sequence of reconstructions that occurs during high-temperature annealing and the steps of the transformation of 3C-SiC(111)/Si(111) toward ML epitaxial graphene by using a combination of high-resolution atomic STM images and DFT calculations. This sequence is expected to be true also for 6H and 4H SiC as the first four layers are the same. The possibility of producing large, continuous domains of epitaxial graphene has been also discussed, as well as the improvements to the graphene quality obtained by substrate polishing.

Acknowledgments

The authors acknowledge the support of the Australian Research Council (ARC) through the Discovery Project DP130102120 and the LIEF grant LE100100146. The authors would like to acknowledge also the support of the Australian National Fabrication Facility and of the Central Analytical Research Facility (CARF), operated by the Institute for Future Environments (QUT). Access to CARF is supported by generous funding from the Science and Engineering Faculty (QUT).

References

1. Novoselov, K., Geim, A. K., Morozov, S., Jiang, D., Zhang, Y., Dubonos, S., Grigorieva, I., and Firsov, A., Electric field effect in atomically thin carbon films, *Science*, **306**(5696), 666–669 (2004).

2. Ferrari, A. C., Bonaccorso, F., Fal'ko, V., Novoselov, K. S., Roche, S., Bøggild, P., Borini, S., Koppens, F. H., Palermo, V., Pugno, N., Garrido, J. A., Sordan, R., Bianco, A., Ballerini, L., Prato, M., Lidorikis, E., Kivioja, J., Marinelli, C., Ryhänen, T., Morpurgo, A., Coleman, J. N., Nicolosi, V., Colombo, L., Fert, A., Garcia-Hernandez, M., Bachtold, A., Schneider, G. F., Guinea, F., Dekker, C., Barbone, M., Sun, Z., Galiotis, C., Grigorenko, A. N., Konstantatos, G., Kis, A., Katsnelson, M., Vandersypen, L., Loiseau, A., Morandi, V., Neumaier, D., Treossi, E., Pellegrini, V., Polini, M., Tredicucci, A., Williams, G. M., Hong, B. H., Ahn, J. H., Kim, J. M., Zirath, H., van Wees, B. J., van der Zant, H., Occhipinti, L., Di Matteo, A., Kinloch, I. A., Seyller, T., Quesnel, E., Feng, X., Teo, K., Rupesinghe, N., Hakonen, P., Neil, S. R., Tannock, Q., Löfwander, T., and Kinaret, J., Science and technology roadmap for graphene, related two-dimensional crystals, and hybrid systems, *Nanoscale*, **7**(11), 4598–4810 (2015).

3. Markoff, J., IBM discloses working version of a much higher-capacity chip, in *The New York Times*, p. B2 (2015).

4. Yoo, E., Kim, J., Hosono, E., Zhou, H.-S., Kudo, T., and Honma, I., Large reversible Li storage of graphene nanosheet families for use in rechargeable lithium ion batteries, *Nano Lett.*, **8**(8), 2277–2282 (2008).

5. Stoller, M. D., Park, S., Zhu, Y., An, J., and Ruoff, R. S., Graphene-based ultracapacitors, *Nano Lett.*, **8**(10), 3498–3502 (2008).

6. Shafiei, M., Spizzirri, P. G., Arsat, R., Yu, J., du Plessis, J., Dubin, S., Kaner, R. B., Kalantar-Zadeh, K., and Wlodarski, W., Platinum/graphene

nanosheet/SiC contacts and their application for hydrogen gas sensing, *J. Phys. Chem. C*, **114**(32), 13796–13801 (2010).

7. Piloto, C., Notarianni, M., Shafiei, M., Taran, E., Galpaya, D., Yan, C., and Motta, N., Highly NO2 sensitive caesium doped graphene oxide conductometric sensors, *Beilstein J. Nanotechnol.*, **5**(1), 1073–1081 (2014).

8. Nayak, T. R., Andersen, H., Makam, V. S., Khaw, C., Bae, S., Xu, X., Ee, P.-L. R., Ahn, J.-H., Hong, B. H., and Pastorin, G., Graphene for controlled and accelerated osteogenic differentiation of human mesenchymal stem cells, *ACS Nano*, **5**(6), 4670–4678 (2011).

9. Wang, X., Zhi, L., and Müllen, K., Transparent, conductive graphene electrodes for dye-sensitized solar cells, *Nano Lett.*, **8**(1), 323–327 (2008).

10. Geim, A. K., and Novoselov, K. S., The rise of graphene, *Nat. Mater.*, **6**(3), 183–191 (2007).

11. Rani, P., and Jindal, V., Designing band gap of graphene by B and N dopant atoms, *RSC Adv.*, **3**(3), 802–812 (2013).

12. Chen, Y.-C., De Oteyza, D. G., Pedramrazi, Z., Chen, C., Fischer, F. R., and Crommie, M. F., Tuning the band gap of graphene nanoribbons synthesized from molecular precursors, *ACS Nano*, **7**(7), 6123–6128 (2013).

13. Son, Y.-W., Cohen, M. L., and Louie, S. G., Energy gaps in graphene nanoribbons, *Phys. Rev. Lett.*, **97**(21), 216803 (2006).

14. Ni, Z. H., Yu, T., Lu, Y. H., Wang, Y. Y., Feng, Y. P., and Shen, Z. X., Uniaxial strain on graphene: Raman spectroscopy study and band-gap opening, *ACS Nano*, **2**(11), 2301–2305 (2008).

15. Zhou, S., Gweon, G.-H., Fedorov, A., First, P., de Heer, W. A., Lee, D.-H., Guinea, F., Neto, A. C., and Lanzara, A., Substrate-induced bandgap opening in epitaxial graphene, *Nat. Mater.*, **6**(10), 770–775 (2007).

16. Severino, A., Bongiorno, C., Piluso, N., Italia, M., Camarda, M., Mauceri, M., Condorelli, G., Di Stefano, M., Cafra, B., and La Magna, A., High-quality 6inch (111) 3C-SiC films grown on off-axis (111) Si substrates, *Thin Solid Films*, **518**(6), S165–S169 (2010).

17. Badami, D. V., Graphitization of alpha-silicon carbide, *Nature*, **193**(4815), 569–570 (1962).

18. Forbeaux, I., Themlin, J. M., Charrier, A., Thibaudau, F., and Debever, J. M., Solid-state graphitization mechanisms of silicon carbide 6H-SiC polar faces, *Appl. Surf. Sci.*, **162**, 406–412 (2000).

19. Ohta, T., Bostwick, A., Seyller, T., Horn, K., and Rotenberg, E., Controlling the electronic structure of bilayer graphene, *Science*, **313**(5789), 951–954 (2006).

20. Berger, C., Song, Z., Li, T., Li, X., Ogbazghi, A. Y., Feng, R., Dai, Z., Marchenkov, A. N., Conrad, E. H., First, P. N., and de Heer, W. A., Ultrathin epitaxial graphite: 2D electron gas properties and a route toward graphene-based nanoelectronics, *J. Phys. Chem. B*, **108**(52), 19912–19916 (2004).

21. Berger, C., Song, Z., Li, X., Wu, X., Brown, N., Naud, C., Mayou, D., Li, T., Hass, J., and Marchenkov, A. N., Electronic confinement and coherence in patterned epitaxial graphene, *Science*, **312**(5777), 1191–1196 (2006).

22. de Heer, W. A., Berger, C., Wu, X., First, P. N., Conrad, E. H., Li, X., Li, T., Sprinkle, M., Hass, J., and Sadowski, M. L., Epitaxial graphene, *Solid State Commun.*, **143**(1), 92–100 (2007).

23. Sutter, P., Epitaxial graphene: how silicon leaves the scene, *Nat. Mater.*, **8**(3), 171–172 (2009).

24. Emtsev, K. V., Bostwick, A., Horn, K., Jobst, J., Kellogg, G. L., Ley, L., McChesney, J. L., Ohta, T., Reshanov, S. A., and Röhrl, J., Towards wafer-size graphene layers by atmospheric pressure graphitization of silicon carbide, *Nat. Mater.*, **8**(3), 203–207 (2009).

25. Ouerghi, A., Silly, M. G., Marangolo, M., Mathieu, C., Eddrief, M., Picher, M., Sirotti, F., El Moussaoui, S., and Belkhou, R., Large-area and high-quality epitaxial graphene on off-axis SiC wafers, *ACS Nano*, **6**(7), 6075–6082 (2012).

26. de Heer, W. A., Berger, C., Ruan, M., Sprinkle, M., Li, X., Hu, Y., Zhang, B., Hankinson, J., and Conrad, E., Large area and structured epitaxial graphene produced by confinement controlled sublimation of silicon carbide, *Proc. Natl. Acad. Sci. U S A*, **108**(41), 16900–16905 (2011).

27. Hass, J., Varchon, F., Millan-Otoya, J. E., Sprinkle, M., Sharma, N., de Heer, W. A., Berger, C., First, P. N., Magaud, L., and Conrad, E. H., Why multilayer graphene on 4H-SiC (0001 over) behaves like a single sheet of graphene, *Phys. Rev. Lett.*, **100**(12), 125504 (2008).

28. Tedesco, J. L., VanMil, B. L., Myers-Ward, R. L., McCrate, J. M., Kitt, S. A., Campbell, P. M., Jernigan, G. G., Culbertson, J. C., Eddy Jr., C., and Gaskill, D. K., Hall effect mobility of epitaxial graphene grown on silicon carbide, *Appl. Phys. Lett.*, **95**, 122102 (2009).

29. Iacopi, F., Walker, G., Wang, L., Malesys, L., Ma, S., Cunning, B. V., and Iacopi, A., Orientation-dependent stress relaxation in hetero-epitaxial 3C-SiC films, *Appl. Phys. Lett.*, **102**(1), 011908-4 (2013).

30. Wang, L., Dimitrijev, S., Han, J., Iacopi, A., Hold, L., Tanner, P., and Harrison, H. B., Growth of 3C–SiC on 150-mm Si (100) substrates by alternating supply epitaxy at 1000° C, *Thin Solid Films*, **519**(19), 6443–6446 (2011).

31. Ouerghi, A., Marangolo, M., Belkhou, R., El Moussaoui, S., Silly, M., Eddrief, M., Largeau, L., Portail, M., Fain, B., and Sirotti, F., Epitaxial graphene on 3C-SiC(111) pseudosubstrate: structural and electronic properties, *Phys. Rev. B*, **82**(12), 125445 (2010).

32. Ouerghi, A., Ridene, M., Balan, A., Belkhou, R., Barbier, A., Gogneau, N., Portail, M., Michon, A., Latil, S., and Jegou, P., Sharp interface in epitaxial graphene layers on 3 C-SiC (100)/Si (100) wafers, *Phys. Rev. B*, **83**(20), 205429 (2011).

33. Aryal, H. R., Fujita, K., Banno, K., and Egawa, T., Epitaxial graphene on Si (111) substrate grown by annealing 3C-SiC/carbonized silicon, *Jpn. J. Appl. Phys.*, **51**(1S), 01AH05 (2012).

34. Ide, T., Kawai, Y., Handa, H., Fukidome, H., Kotsugi, M., Ohkochi, T., Enta, Y., Kinoshita, T., Yoshigoe, A., and Teraoka, Y., Epitaxy of graphene on 3C-SiC (111) thin films on microfabricated Si (111) substrates, *Jpn. J. Appl. Phys.*, **51**(6S), 06FD02 (2012).

35. Takahashi, R., Handa, H., Abe, S., Imaizumi, K., Fukidome, H., Yoshigoe, A., Teraoka, Y., and Suemitsu, M., Low-energy-electron-diffraction and X-ray-phototelectron-spectroscopy studies of graphitization of 3C-SiC (111) thin film on Si (111) substrate, *Jpn. J. Appl. Phys.*, **50**(7), 0103 (2011).

36. Suemitsu, M., and Fukidome, H., Epitaxial graphene on silicon substrates, *J. Phys. D: Appl. Phys.*, **43**(37), 374012 (2010).

37. Chaika, A. N., Molodtsova, O. V., Zakharov, A. A., Marchenko, D., Sánchez-Barriga, J., Varykhalov, A., Shvets, I. V., and Aristov, V. Y., Continuous wafer-scale graphene on cubic-SiC (001), *Nano Res.*, **6**(8), 562–570 (2013).

38. Aristov, V. Y., Urbanik, G., Kummer, K., Vyalikh, D. V., Molodtsova, O. V., Preobrajenski, A. B., Zakharov, A. A., Hess, C., Hänke, T., and Büchner, B., Graphene synthesis on cubic SiC/Si wafers. Perspectives for mass production of graphene-based electronic devices, *Nano Lett.*, **10**(3), 992–995 (2010).

39. Ouerghi, A., Balan, A., Castelli, C., Picher, M., Belkhou, R., Eddrief, M., Silly, M., Marangolo, M., Shukla, A., and Sirotti, F., Epitaxial graphene on single domain 3C-SiC (100) thin films grown on off-axis Si (100), *Appl. Phys. Lett.*, **101**(2), 21603 (2012).

40. Ouerghi, A., Belkhou, R., Marangolo, M., Silly, M., El Moussaoui, S., Eddrief, M., Largeau, L., Portail, M., and Sirotti, F., Structural coherency of epitaxial graphene on 3C–SiC (111) epilayers on Si (111), *Appl. Phys. Lett.*, **97**, 161905 (2010).

41. Starke, U., Coletti, C., Emtsev, K., Zakharov, A. A., Ouisse, T., and Chaussende, D., Large area quasi-free standing monolayer graphene on 3C-SiC (111), *Mater. Sci. Forum*, **717–720**, 617–620 (2012).

42. Darakchieva, V., Boosalis, A., Zakharov, A., Hofmann, T., Schubert, M., Tiwald, T., Iakimov, T., Vasiliauskas, R., and Yakimova, R., Large-area microfocal spectroscopic ellipsometry mapping of thickness and electronic properties of epitaxial graphene on Si-and C-face of 3C-SiC (111), *Appl. Phys. Lett.*, **102**(21), 213116 (2013).

43. Ouerghi, A., Kahouli, A., Lucot, D., Portail, M., Travers, L., Gierak, J., Penuelas, J., Jegou, P., Shukla, A., and Chassagne, T., Epitaxial graphene on cubic SiC (111)/Si (111) substrate, *Appl. Phys. Lett.*, **96**(19), 191910-191910-3 (2010).

44. Yazdi, G. R., Vasiliauskas, R., Iakimov, T., Zakharov, A., Syväjärvi, M., and Yakimova, R., Growth of large area monolayer graphene on 3C-SiC and a comparison with other SiC polytypes, *Carbon*, **57**, 477–484 (2013).

45. Pierucci, D., Sediri, H., Hajlaoui, M., Girard, J.-C., Brumme, T., Calandra, M., Velez-Fort, E., Patriarche, G., Silly, M. G., Ferro, G., Soulière, V., Marangolo, M., Sirotti, F., Mauri, F., and Ouerghi, A., Evidence for flat bands near the Fermi level in epitaxial rhombohedral multilayer graphene, *ACS Nano*, **9**(5), 5432–5439 (2015).

46. Coletti, C., Forti, S., Principi, A., Emtsev, K. V., Zakharov, A. A., Daniels, K. M., Daas, B. K., Chandrashekhar, M. V. S., Ouisse, T., Chaussende, D., MacDonald, A. H., Polini, M., and Starke, U., Revealing the electronic band structure of trilayer graphene on SiC: an angle-resolved photoemission study, *Phys. Rev. B*, **88**(15), 155439 (2013).

47. Gupta, B., Placidi, E., Hogan, C., Mishra, N., Iacopi, F., and Motta, N., The transition from 3C SiC(111) to graphene captured by ultra high vacuum scanning tunneling microscopy, *Carbon*, **91**, 378–385 (2015).

48. Gupta, B., Notarianni, M., Mishra, N., Shafiei, M., Iacopi, F., and Motta, N., Evolution of epitaxial graphene layers on 3C SiC/Si (111) as a function of annealing temperature in UHV, *Carbon*, **68**, 563–572 (2014).

49. Van Bommel, A., Crombeen, J., and Van Tooren, A., LEED and Auger electron observations of the SiC (0001) surface, *Surf. Sci.*, **48**(2), 463–472 (1975).

50. Starke, U., Schardt, J., Bernhardt, J., Franke, M., Reuter, K., Wedler, H., Heinz, K., Furthmüller, J., Käckell, P., and Bechstedt, F., Novel reconstruction mechanism for dangling-bond minimization: combined method surface structure determination of SiC (111)-(3× 3), *Phys. Rev. Lett.*, **80**(4), 758 (1998).

51. Riedl, C., Coletti, C., and Starke, U., Structural and electronic properties of epitaxial graphene on SiC (0 0 0 1): a review of growth, characterization, transfer doping and hydrogen intercalation, *J. Phys. D: Appl. Phys.*, **43**(37), 374009 (2010).

52. Northrup, J. E., and Neugebauer, J., Theory of the adatom-induced reconstruction of the SiC(0001)$\sqrt{3}\times\sqrt{3}$ surface, *Phys. Rev. B*, **52**(24), R17001–R17004 (1995).

53. Owman, F., and Mårtensson, P., STM study of the SiC (0001)$\sqrt{3} \times \sqrt{3}$ surface, *Surf. Sci.*, **330**(1), L639–L645 (1995).

54. Sabisch, M., Krüger, P., and Pollmann, J., Ab initio calculations of structural and electronic properties of 6H-SiC (0001) surfaces, *Phys. Rev. B*, **55**(16), 10561 (1997).

55. Li, L., and Tsong, I., Atomic structures of 6H SiC (0001) and (000$\bar{1}$) surfaces, *Surf. Sci.*, **351**(1), 141–148 (1996).

56. Starke, U., and Riedl, C., Epitaxial graphene on SiC(0001) and SiC(000$\bar{1}$): from surface reconstructions to carbon electronics, *J. Phys.: Condens. Matter*, **21**(13), 134016 (2009).

57. Lauffer, P., Emtsev, K., Graupner, R., Seyller, T., Ley, L., Reshanov, S., and Weber, H., Atomic and electronic structure of few-layer graphene on SiC (0001) studied with scanning tunneling microscopy and spectroscopy, *Phys. Rev. B*, **77**(15), 155426 (2008).

58. Schardt, J., Bernhardt, J., Starke, U., and Heinz, K., Crystallography of the (3× 3) surface reconstruction of 3 C-SiC (111), 4 H-SiC (0001), and 6 H-SiC (0001) surfaces retrieved by low-energy electron diffraction, *Phys. Rev. B*, **62**(15), 10335 (2000).

59. Starke, U., Schardt, J., and Franke, M., Morphology, bond saturation and reconstruction of hexagonal SiC surfaces, *Appl. Phys. A*, **65**(6), 587–596 (1997).

60. Dimitrakopoulos, C., Lin, Y.-M., Grill, A., Farmer, D. B., Freitag, M., Sun, Y., Han, S.-J., Chen, Z., Jenkins, K. A., and Zhu, Y., Wafer-scale epitaxial graphene growth on the Si-face of hexagonal SiC (0001) for high frequency transistors, *J. Vac. Sci. Technol. B*, **28**(5), 985–992 (2010).

61. Hass, J., Millán-Otoya, J., First, P., and Conrad, E., Interface structure of epitaxial graphene grown on 4H-SiC (0001), *Phys. Rev. B*, **78**(20), 205424 (2008).

62. Varchon, F., Feng, R., Hass, J., Li, X., Nguyen, B. N., Naud, C., Mallet, P., Veuillen, J. Y., Berger, C., Conrad, E. H., and Magaud, L., Electronic structure of epitaxial graphene layers on SiC: effect of the substrate, *Phys. Rev. Lett.*, **99**(12), 126805 (2007).

63. Emtsev, K., Speck, F., Seyller, T., Ley, L., and Riley, J., Interaction, growth, and ordering of epitaxial graphene on SiC {0001} surfaces: a comparative photoelectron spectroscopy study, *Phys. Rev. B*, **77**(15), 155303 (2008).

64. Hiebel, F., Mallet, P., Varchon, F., Magaud, L., and Veuillen, J., Graphene-substrate interaction on 6 H-SiC (000$\bar{1}$): a scanning tunneling microscopy study, *Phys. Rev. B*, **78**(15), 153412 (2008).

65. Hibino, H., Kageshima, H., and Nagase, M., Epitaxial few-layer graphene: towards single crystal growth, *J. Phys. D: Appl. Phys.*, **43**(37), 374005 (2010).

66. Magaud, L., Hiebel, F., Varchon, F., Mallet, P., and Veuillen, J.-Y., Graphene on the C-terminated SiC (000 1) surface: an ab initio study, *Phys. Rev. B*, **79**(16), 161405 (2009).

67. Avouris, P., and Dimitrakopoulos, C., Graphene: synthesis and applications, *Mater. Today*, **15**(3), 86–97 (2012).

68. Varchon, F., Mallet, P., Magaud, L., and Veuillen, J.-Y., Rotational disorder in few-layer graphene films on 6 H-Si C (000-1): a scanning tunneling microscopy study, *Phys. Rev. B*, **77**(16), 165415 (2008).

69. Abe, S., Handa, H., Takahashi, R., Imaizumi, K., Fukidome, H., and Suemitsu, M., Surface chemistry involved in epitaxy of graphene on 3C-SiC (111)/Si (111), *Nanoscale Res. Lett.*, **5**(12), 1888–1891 (2010).

70. Suemitsu, M., Miyamoto, Y., Handa, H., and Konno, A., Graphene formation on a 3C-SiC (111) thin film grown on Si (110) substrate, *e-J. Surf. Sci. Nanotechnol.*, **7**, 311–313 (2009).

71. Mårtensson, P., Owman, F., and Johansson, L., Morphology, atomic and electronic structure of 6H-SiC (0001) surfaces, *Phys. Status Solidi B*, **202**(1), 501–528 (1997).

72. Fukidome, H., Abe, S., Takahashi, R., Imaizumi, K., Inomata, S., Handa, H., Saito, E., Enta, Y., Yoshigoe, A., and Teraoka, Y., Controls over structural and electronic properties of epitaxial graphene on silicon using surface termination of 3C-SiC (111)/Si, *Appl. Phys. Express*, **4**(11), 115104 (2011).

73. Hass, J., de Heer, W. A., and Conrad, E., The growth and morphology of epitaxial multilayer graphene, *J. Phys.: Condens. Matter*, **20**(32), 323202 (2008).

74. Giannozzi, P., Baroni, S., Bonini, N., Calandra, M., Car, R., Cavazzoni, C., Ceresoli, D., Chiarotti, G.L., Cococcioni, M., and Dabo, I., QUANTUM ESPRESSO: a modular and open-source software project for quantum simulations of materials, *J. Phys.: Condens. Matter*, **21**(39), 395502 (2009).

75. Tersoff, J., and Hamann, D. R., Theory of the scanning tunneling microscope, *Phys. Rev. B*, **31**, 805–813 (1985).

76. Wong, S. L., Huang, H., Chen, W., and Wee, A. T. S., STM studies of epitaxial graphene, *MRS Bull.*, **37**(12), 1195–1202 (2012).

77. Gupta, B., Bernardo, I. D., Mondelli, P., Pia, A. D., Betti, M. G., Iacopi, F., Mariani, C., and Motta, N., Effect of substrate polishing on the growth of graphene on 3C–SiC(111)/Si(111) by high temperature annealing, *Nanotechnology*, **27**(18), 185601 (2016).

78. Ohta, T., Bostwick, A., McChesney, J. L., Seyller, T., Horn, K., and Rotenberg, E., Interlayer interaction and electronic screening in multilayer graphene investigated with angle-resolved photoemission spectroscopy, *Phys. Rev. Lett.*, **98**(20), 206802 (2007).

79. Johansson, L. I., Armiento, R., Avila, J., Xia, C., Lorcy, S., Abrikosov, I. A., Asensio, M. C., and Virojanadara, C., Multiple π-bands and Bernal stacking of multilayer graphene on C-face SiC, revealed by nano-angle resolved photoemission, *Sci. Rep.*, **4**, 4157 (2014).

Chapter 4

Diffusion and Kinetics in Epitaxial Graphene Growth on SiC

M. Tomellini,[a] B. Gupta,[b] A. Sgarlata,[c] and N. Motta[b]

[a]*Dipartimento di Scienze e Tecnologie Chimiche, Università degli Studi di Roma Tor Vergata,*
Via Della Ricerca Scientifica 1, 00133 Roma, Italy
[b]*School of Chemistry, Physics and Mechanical Engineering and Institute for Future Environments, Queensland University of Technology, 2 George Street, Brisbane 4001, QLD, Australia*
[c]*Dipartimento di Fisica, Università degli Studi di Roma Tor Vergata,*
Via Della Ricerca Scientifica 1, 00133 Roma, Italy
n.motta@qut.edu.au

4.1 Introduction

One of the problems that graphene is facing in electronic applications is its quality, which is still far from the level required to obtain an industrial-scale production of reliable nanoscale devices, and it is ultimately related to the growth method. Understanding the kinetics of graphene growth is an outstanding problem in physics, whose solution is the key to achieving full control on graphene's properties.

Growing Graphene on Semiconductors
Edited by Nunzio Motta, Francesca Iacopi, and Camilla Coletti
Copyright © 2017 Pan Stanford Publishing Pte. Ltd.
ISBN 978-981-4774-21-5 (Hardcover), 978-1-315-18615-3 (eBook)
www.panstanford.com

This chapter deals with the microscopic phenomena occurring during epitaxial graphene growth on SiC by high-temperature annealing. We discuss recent experimental and theoretical findings on the Si diffusion in the SiC matrix and on the kinetics of graphene growth as a function of temperature and time. The theoretical framework developed in this chapter will help to understand the physical basis of graphene formation, which is essential for perfect control of graphene quality, helping to define the optimal conditions for the growth of continuous monolayer or bilayer graphene on SiC.

In the last few years several authors demonstrated the possibility of growing perfectly controlled layers of graphene on SiC(0001) by using thermal annealing in a furnace under an Ar atmosphere [1], in a pressurized vessel [2], or in ultrahigh vacuum (UHV) under Si flux and N_2 partial pressure [3]. All these studies agree on the fact that Si sublimates too quickly in UHV. Thermal annealing of SiC in UHV, at temperatures where Si diffuses out from the surface, occurs in a nonequilibrium condition. To bring the system as close as possible to the equilibrium several strategies have been employed to limit Si sublimation, obtaining large, continuous domains of high-quality graphene films on SiC [4]. A number of studies have been devoted to the graphitization and formation of the first layer [5–20], which occurs passing through several reconstructions.

Sun et al. [21] analyzed with the help of first principles calculations the diffusion of Si atoms through heptagon-pentagon defects of the $(6\sqrt{3}\times6\sqrt{3})R30°$ (defined $6\sqrt{3}$ in the following), demonstrating that after the formation of this structure, entirely made by C atoms, Si is no more exposed for an easy desorption. So the Si out-diffusion becomes the rate-limiting step. They suggest two alternative pathways for the diffusion: vertically, through the $6\sqrt{3}$ defects, and laterally, via a step edge. Huang et al. and Hannon et al. [22, 23] suggested a bottom-up growth mode epitaxial graphene on SiC in UHV where upon annealing at a high temperature, the interface layer $6\sqrt{3}$ develops first, it then turns into graphene, and then another interface layer is created underneath.

Tanaka et al. [24] and Drabińska at al. [25] reported a power law dependence of the number of graphene layers as a function of time, but they did not provide a full theory of the phenomenon.

As a matter of fact, a detailed theory of the epitaxial graphene formation by SiC thermal decomposition beyond the first layer is missing, and therefore it would be essential to develop a physical model to gain insight and achieve perfect control on the phenomenon.

In the following we present the most recent results obtained in the growth of graphene on SiC, introducing a full model that accounts for the growth mechanism of multilayer epitaxial graphene by high-temperature annealing of SiC. Some of the studies we discuss in this chapter have been performed on SiC(111)/Si(111) as this material is the basis for application of graphene in Si nanoelectronics as well as a way to overcome the high cost of bulk SiC as a substrate for graphene growth. However these results are equally applicable to the growth of graphene on SiC(0001) as the first four layers are the same.

4.2 Evolution of Epitaxial Graphene Films as a Function of Annealing Temperature

Among the first investigations of graphene growth on SiC as a function of annealing temperature [7, 8, 11] we discuss the one performed by Ouerghi et al. [11] on the Si-terminated face of 3C SiC(111)/Si(111). Under UHV conditions they firstly degassed the substrate at 600°C, followed by annealing under low Si flux at 900°C to remove the native oxide from the surface. The graphene growth was carried by annealing the sample for 10 min at a temperature (T) ranging from 900°C to 1300°C. After each annealing a high-resolution X-ray photoelectron spectroscopy (XPS) spectrum around the carbon (C) 1s peak (binding energy, BE $\approx$284.4 eV) was acquired (Fig. 4.1a). It is possible to recognize three contributions in C 1s: a C-Si peak at ~283.4 eV coming from SiC, a C–C peak at ~284.6 eV coming from graphene, and a peak at ~285.5 eV coming from C in the interface layer. The authors found that on increasing the temperature the graphene peak (C–C) undergoes a shift while its intensity increases (Fig. 4.1b).

A more elaborate and detailed discussion of the evolution of the graphene (C–C) peak at each temperature has been performed by Gupta et al. [26], providing a better understanding of the growth process [27, 28]. After annealing at 950°C under Si flux, the formation of a Si-rich (3 × 3) surface is observed [29]. A comparison between the high-resolution XPS spectra obtained at 650°C and 950°C for both C 1s (283.1 eV) and Si 2p (100.9 eV) peaks is shown in Fig. 4.2. After 950°C annealing, it is evident that there is a slight increase (0.75 versus 0.68) in the intensity ratio of the Si 2p peak with respect to the C 1s and the two side peaks in the Si 2p disappear, confirming the effect of the Si evaporation in removing the SiO_2 and in increasing the Si/C ratio at the surface.

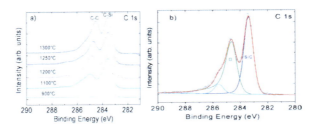

Figure 4.1 (a) C 1s XPS spectra of fully grown graphene on 3C SiC(111) for different temperatures. (b) XPS spectra of the C 1s core level for graphene with a Doniach–Sunjic line shape analysis (red line). Reprinted from Ref. [11], with the permission of AIP Publishing.

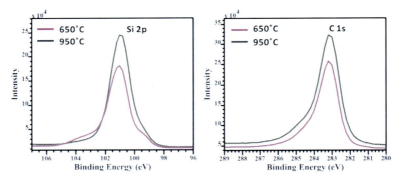

Figure 4.2 X-ray photoelectron spectroscopy of the SiC(111) Si-rich surface after 650°C and 950°C annealing with Si flux: (a) Si 2p and (b) C 1s [26].

The evolution of the C 1s peak after 10 min annealing at increasing temperatures is presented in Fig. 4.3. The peaks used to fit the original spectra of C 1s are SiC (~283 eV), graphene (G = 284.7 eV), and buffer layer (I = 285.85 eV), which are in agreement with what was reported for hexagonal SiC polytypes (i.e., 4H and 6H SiC) [8, 10]. At 1125°C (Fig. 4.3a), the component at 285.85 eV, identified

Evolution of Epitaxial Graphene Films as a Function of Annealing Temperature | 113

as the buffer layer peak, begins to develop aside the main SiC bulk peak at 283.0 eV. This is attributed to the transition from a Si-rich surface to a C-rich surface [29]. This layer has a typical $(6\sqrt{3} \times 6\sqrt{3})$ $R30°$ reconstruction, as proven by STM images and XPS spectra in the recent literature. The third peak, at 284.7 eV, is assigned to graphene, as it is only 0.3 eV behind the typical graphitic carbon peak. Around 1225°C (Fig. 4.3b), the I and G components start to increase.

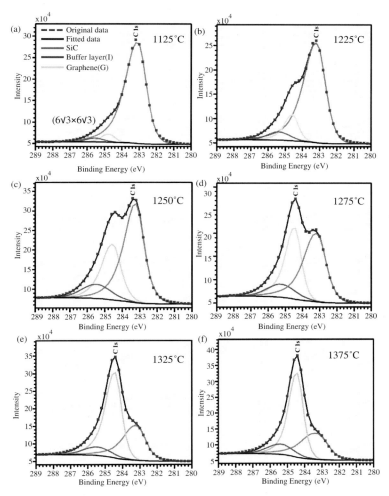

Figure 4.3 C 1s XPS spectra for graphene growth after 10 min annealing as a function of increasing temperature. The sample was annealed at (a) 1125°C, (b) 1225°C, (c) 1250°C, (d) 1275°C, (e) 1325°C and (f) 1375°C [26].

Diffusion and Kinetics in Epitaxial Graphene Growth on SiC

The increase of the graphene peak intensity and decrease of the SiC peak intensity (Fig. 4.3c–f) indicates the formation of an increasing number of graphene layers as a function of annealing temperature. The number of graphene layers at each temperature was calculated by using the intensity ratio of graphene (N_G) and the SiC peak as a reference (N_R) from XPS, obtained by the formula [30]

$$\frac{N_G}{N_R} = \frac{T(E_G)\varrho'C_G\lambda'(E_G)\left[1-\exp\left(-\dfrac{t}{\lambda'(E_G)}\right)\right]}{T(E_R)\rho C_R\lambda(E_R)\exp\left(-\dfrac{t}{\lambda'(E_R)}\right)}\cdot F\,,\qquad(4.1)$$

where E is the kinetic energy of photoelectrons associated with a given peak, T is the transmission function of the analyzer, C is the differential cross section ($d\sigma/d\Omega$), and ρ and λ are, respectively, the atomic density and inelastic mean free path [30] of the corresponding material. F is a geometrical correction factor due to photoelectron diffraction and the $'$ indicates quantities referred to the graphene overlayer as opposed to the SiC bulk.

4.2.1   Evaluation of the Growth Rate in UHV

The growth rate as a function of temperature (Fig. 4.4) was obtained by calculating the number of layers developed after 10 min annealing from Eq. 4.1, as a function of annealing temperature, and fitting the data by using the Arrhenius formula

$$N(T) = Ae^{E_a/k_BT}\,.\qquad(4.2)$$

From this fitting an activation energy of E_a = 2.46 eV was calculated [26, 31].

Abe et al. [32] also investigated the growth of graphene on 3C-SiC(111)/Si(111) in UHV. They annealed 3C SiC thin films, with a thickness of ~100 nm at three different temperatures: 1000°C, 1200°C, and 1300°C. They found that the epitaxy of graphene on 3C SiC(111)/Si(111) proceeds in a similar manner to that on a hexagonal SiC(0001) bulk crystal.

Fukidome et al. [33] studied the epitaxial graphene growth on both the Si and C faces of 3C SiC(111). They showed how the surface termination of 3C SiC(111)/Si affects the stacking, interface structure, and electronic properties of graphene on 3C SiC. For the Si face, they used 3C SiC(111)/Si(111) and for the C face, a 3C SiC(111)/

Si(110) (rotated epitaxy) surface. The epitaxial graphene was grown on both surfaces at 1250°C for 30 min in vacuum. The peak due the buffer layer was only observed on the 3C SiC(111)/Si(111) by the C 1s core-level spectra (Fig. 4.5a).

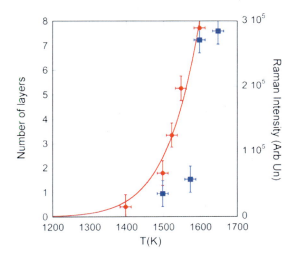

Figure 4.4 Number of graphene layers developed in 10 min versus the annealing temperature, as obtained from XPS analysis (red dots). The values are fitted to the Arrhenius function shown in Eq. 4.2 (solid line). Blue squares: area of 2D Raman peak versus the annealing temperature [26].

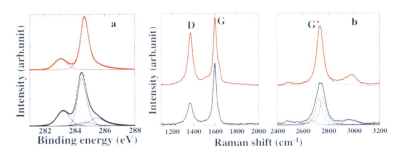

Figure 4.5 Characterization of epitaxial graphene on nonrotated 3C SiC(111)/Si(111) (black) and rotated 3C SiC(111)/Si(110) (red). (a) C 1s core level; spectra of the epitaxial graphene on 3C SiC(111)/Si(111) and 3C SiC(111)/Si(110). The incident photon energy is 650 eV, and the energy resolution is <20 meV. (b) Raman spectra of the epitaxial graphene on 3C SiC(111)/Si(111) and 3C SiC(111)/Si(110). The laser energy for the excitation is 2.41 eV. Adapted from Ref. [33]. Copyright (2011) The Japan Society of Applied Physics.

4.3 Growth Kinetics of Epitaxial Graphene Films on SiC

All recent studies on graphene growth by sublimation of SiC agree on the fact that Si sublimates too quickly in UHV [22, 24, 25, 34]. Thermal annealing of SiC in UHV, at temperatures where Si diffuses out from the surface, occurs in a nonequilibrium condition. To bring the system as close as possible to the equilibrium several strategies [2, 3] have been employed to limit the Si sublimation, obtaining large, continuous $6\sqrt{3}\times6\sqrt{3}$ domains on SiC, which enable the subsequent formation of high-quality graphene films [25]. A number of studies have been devoted to the graphitization and formation of the first layer [24] as well as to the growth mechanism and kinetics of the process [26, 35].

4.3.1 Growth Kinetics under Ar Pressure

Drabinska et al. [25] studied the growth kinetics of epitaxial graphene on on-axis SiC substrates in a chemical vapor deposition (CVD) reactor by varying the argon pressure and the annealing time. They used optical absorption in order to calculate the number of layers of graphene.

The curve in Fig. 4.6a shows a decrease of optical absorption and of the number of graphene layers as a function of increasing argon pressure. The authors report that for a viscous flow regime, corresponding to their experimental conditions, the evaporation rate of the Si atom from the sample surface should be inversely proportional to the argon pressure, although they observed a slight deviation from this law. They fitted the data by using the formula for the number of layers $N = \dfrac{C}{p + p_0}$ with $p_0 = 48 \pm 9$ mbar. p_0 is assigned to an "effective pressure" of Si vapor interacting with graphene layers and/or graphene-SiC interface. The optical absorption increases with annealing time (Fig. 4.6b), showing a nonlinear behavior, which was fitted to a power law with an exponent equal to 0.35. To explain this result they argue that Si atoms cannot escape easily through already grown graphene layers, so the process of Si out-diffusion should be more complex than in common cases like iron or Si oxidation at a high temperature, where the exponent $d \approx 0.5$ is observed. They conclude that Si atom evaporation and 2D interlayer diffusion should be responsible for the growth kinetics of multilayer graphene structures on SiC. However, they did not comment about the quality of graphene obtained by increasing the annealing time.

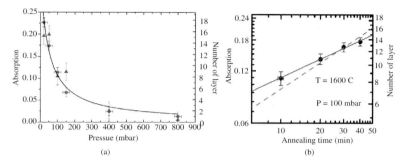

Figure 4.6 (a) The optical absorption dependence on argon pressure during growth process at a temperature T = 1600°C and an annealing time t = 10 min. (b) The optical absorption dependence on annealing time in growth process at a temperature T = 1600°C and argon pressure p = 100 mbar. Dashed line and solid line: power law dependence with the exponents 0.5 and 0.35, respectively. Adapted (figure) with permission from Ref. [25]. Copyright (2010) by the American Physical Society.

4.3.2 Growth Kinetics in UHV

Tanaka et al. [24] used vicinal SiC substrates in order to grow graphene in UHV, finding that the quality of the graphene layers was better compared to the ones grown on on-axis substrates. In their experiment the graphene growth proceeds in layer-by-layer mode along the step edges, which has been identified as anisotropic layer-by-layer growth [24]. The layer thickness increases nonlinearly with annealing time (Fig. 4.7), with a very steep growth rate in the initial stage (up to a thickness of ~1 ML), which reduces consistently in the stages that follow. This pattern suggests that the growth rate is limited by the graphene layer thickness. According to this, they suggested a schematic process of graphene growth, including three stages, as shown in Fig. 4.8. Stage I (0–1 ML) consists of the formation of the buffer layer and nucleation of 1 ML graphene at the nanofacets, followed by 1 ML graphene growth over the entire surface. Stage II (1–2 ML) consists of the nucleation of the second graphene layer at the nanofacets below the first graphene layer and a continuous graphene growth via layer-by-layer mode. Stage III (2–3 ML) is similar to Stage II in all respects except that the graphene thickness has increased. The growth rate decreases with the increase of graphene thickness, which is indicative of a Si desorption (out-diffusion) limited process. They maintain that the graphene layer acts as a Si-diffusion barrier, creating a thermal equilibrium condition, which is probably the key factor in achieving graphene layers with a high crystal quality [24, 34].

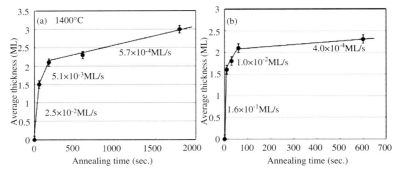

Figure 4.7 Average layer thickness of graphene for various annealing time lengths at (a) 1400°C and (b) 1600°C. The stepwise decrease in growth rate with the increase in the graphene layer thickness [24].

4.3.3 Si and C Diffusion Process

The key to the formation of the graphene layer is in the microscopic details of the diffusion process, which were recently analyzed by Sun et al. [21]. They studied the diffusion of Si atoms through heptagon-pentagon defects of the $6\sqrt{3} \times 6\sqrt{3}$, demonstrating that after the formation of this structure, entirely made by C atoms, Si is no more exposed for an easy desorption.

As Si out-diffusion becomes the rate-limiting step, they suggest two alternative pathways: vertical diffusion through the $6\sqrt{3}$ defects and lateral diffusion via a step edge. Figure 4.9 shows the vertical diffusion pathway through the $6\sqrt{3}$ layer via first principle calculations, indicating an energetically competitive path for the out-diffusion of Si.

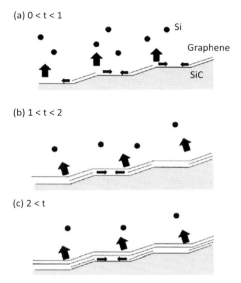

Figure 4.8 Schematic models of graphene growth at each stage. (a) Stage I: 0 to 1 ML via the $(6\sqrt{3} \times 6\sqrt{3})R30°$ buffer layer. (b) Stage II: 1 to 2 ML. (c) Stage III: 2 to 3 ML. An overgrown graphene layer acts as a diffusion barrier for the Si desorption, resulting in the decrease in growth rates [24].

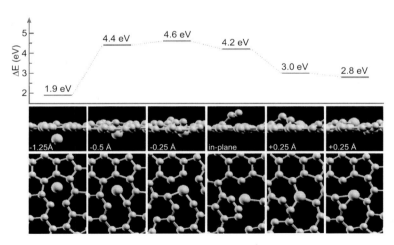

Figure 4.9 (Top panel) Energies and (middle panel) snapshots of the vertical migration process of Si through the $6\sqrt{3}$ layer in cross-section and (bottom panel) plane view. Reprinted (figure) with permission from Ref. [21]. Copyright (2011) by the American Physical Society.

Huang et al. suggested a bottom-up growth mode for epitaxial graphene on SiC in UHV, where upon annealing at a high temperature the interface 6√3 layer develops first, it then turns into graphene, and then another interface layer is created underneath [22].

Pham et al. [34] studied the Si out-diffusion in UHV from the 3C SiC(111)/Si substrate during graphene growth by C evaporation on Si. Their method involves the growth of a 3C SiC/Si layer by coevaporation of Si and C in UHV at 1000°C followed by evaporation of C at 1200°C. They measured the atomic concentration of Si by an XPS depth profile before and after graphene growth. The concentration of Si and C atoms were nearly constant before graphene growth. After annealing the sample at 1100°C for 2 h, they found that the Si concentration has slightly increased (Fig. 4.10a). This was due to the out-diffusion of Si from the substrate during annealing [36, 37]. In Fig. 4.10b they show a comparison between measured atomic concentrations of Si and calculation based on diffusion equation, where they evaluate the diffusion coefficient by a best-fitting procedure. The diffusion coefficient was larger than that evaluated by other authors, and they have assigned it to the presence of more defects on their substrate and maybe error in surface temperature measurement. They also found that Si flux is always present during carbon growth due to Si out-diffusion from the surface through SiC layers at a high temperature. Finally they reported that a high temperature can help to reduce the surface roughness but is harmful for the quality of graphene as it stimulates Si out-diffusion from the substrate and intermixing with the deposited carbon at the surface.

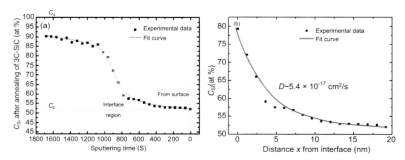

Figure 4.10 (a) Concentration of Si atoms C_{Si} versus sputtering time from the sample surface after annealing ~20 nm thick 3C SiC on Si(111) at 1100°C for 2h. (b) Measured Si concentration profile for determining the diffusion coefficient D of Si. Reprinted from Ref. [34], Copyright (2016), with permission from Elsevier.

4.3.4 Kinetic Model of Graphene Layer-by-Layer Formation

Recently Zarotti et al. [35] presented an extensive study of the kinetics of epitaxial graphene growth. They monitored the intensity of the XPS C 1s peak as a function of time and of annealing temperature. The relative intensities of the three components of the C 1s peak (SiC, graphene, and interface layer, as discussed before) change during the annealing, revealing the evolution of the surface composition as a function of annealing time (Fig. 4.11a,b). The number of layers grows with time, with a law that depends on the temperature (Fig. 4.11c). At the highest T (1325°C) the number of layers after 1800 s is about nine.

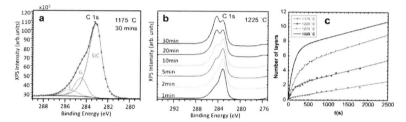

Figure 4.11 (a) Fitting of an XPS C 1s peak with three different components, identified as I (interface layer), G (graphene), and SiC. The spectrum refers to the growth of graphene for 30 min at 1175°C. (b) Time evolution of C 1s spectra for graphene growth at 1225°C [35]. (c) Number of layers as a function of time and temperature, obtained by using the formula in Eq. 4.1.

To model the growth one simplifying assumption was made in dealing with the intensity of the C 1s components: the SiC peak was defined as a "substrate" component (I_{sub}), and the sum of G and I peaks as "overlayer" component (I_{ov}). By considering a bottom-up mechanism for the development of the graphene layers, where Si atoms leave the buffer layer and C atoms rearrange in a new 2D network (Fig. 4.12) [36, 37], a layer-by-layer (Frank-van der Merwe) growth is expected.

The XPS intensity signals for the overlayer ($I_{ov}(t)$) and the substrate ($I_{sub}(t)$) are expressed as a function of t:

$$\begin{cases} I_{ov}(t) = I_{ov}^{\infty}[1 - e^{-n(t)}] \\ I_{sub}(t) = I_{sub}^{0} e^{-n(t)} \end{cases} \quad (4.3)$$

by defining a dimensionless function

$$n(t) = h(t)/\lambda \tag{4.4}$$

as the ratio between $h(t)$, the mean overlayer thickness that forms on the substrate at the time t, and the effective escape depth of electrons in the material λ ($\lambda = 28$ Å for graphene [30, 35]. In this expression I_{ov}^{∞} is the asymptotic value for the overlayer intensity (unknown) and I_{sub}^{0} is the known value for the substrate intensity at $t = 0$. Zarotti et al. [35] used normalized expressions:

$$\tilde{I}_{ov}(t) = \frac{I_{ov}^{\infty}}{\tilde{I}_{ov}^{0}}[1 - e^{-n(t)}] \tag{4.5}$$

$$\tilde{I}_{sub}(t) = \frac{I_{sub}^{0}}{I_{SiC}^{max}}e^{-n(t)}, \tag{4.6}$$

where I_{sub}^{0} is a constant value equal to $2 \cdot 10^5$ and I_{SiC}^{max}, which is temperature dependent, is the maximum value of the experimental SiC component. Note that $\dfrac{I_{sub}^{0}}{I_{SiC}^{max}} \approx 1$ as the substrate is SiC and the maximum intensity of the SiC contribution is obtained at $t = 0$. The relation

$$\tilde{I}_{ov}(t) \approx m[1 - \tilde{I}_{sub}(t)] \tag{4.7}$$

that can be obtained by combining Eqs. 4.5 and 4.6 is necessary to evaluate the asymptotic value I_{ov}^{∞} for each temperature. In fact by plotting the overlayer measured intensity versus the substrate intensity (see the inset of Fig. 4.13b), it is possible to evaluate the slope m of Eq. 4.7, determining the unknown asymptotic value I_{ov}^{∞} for each temperature. By using the above values as parameters of Eq. 4.3, a fitting of the overlayer data was obtained, as explained in Ref. [35] in the hypothesis of $n(t) = \beta t^{\gamma}$, leading to $\gamma = 1/2$ as the common parameter providing the best fit at all temperatures. The formula

$$n(t) = \beta\sqrt{t} \tag{4.8}$$

provides also the value of the prefactor β at each temperature, where $1/\beta^2$ is the characteristic time of the growth process. The results of the fits are displayed in Fig. 4.13a,b, and the values of $1/\beta^2$ and of the maximum thickness are reported in Table 4.1.

Growth Kinetics of Epitaxial Graphene Films on SiC | 123

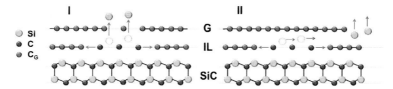

Figure 4.12 Model for epitaxial graphene formation on SiC in the layer-by-layer approximation. Left: Si atoms leave the substrate from defects. Right: Si atoms need to travel along the surface to reach a step edge, from where they leave [35].

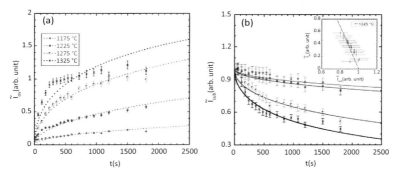

Figure 4.13 Evolution of (a) the overlayer and of (b) the substrate intensity as a function of time at different annealing temperatures along with fitting using Eqs. 4.5 and 4.6, where $n(t) = \beta\sqrt{t}$. The details of the fitting procedure are explained in the text. The inset shows the linear relationship between the overlayer and substrate intensity as a function of time at 1225°C, providing the linear relationship in Eq. 4.7 necessary to evaluate the asymptotic value I_{ov}^{∞}.

Table 4.1 Values of the calculated $1/\beta^2$ along with the maximum thickness of the graphene layer, obtained at t = 1800 s in Ref. [35]

T (°C)	1175	1225	1275	1325
$1/\beta^2 (s)$	69 10³	46 10³	5.3 10³	2.3 10³
$h(t_{max})$ (Å)	4.5	5.6	16	21

The agreement between the fitting and the experimental data is good for most temperatures. The small deviation of the fitting from the data at 1325°C suggests that at this temperature some other mechanism becomes predominant in the graphene formation. This mechanism is discussed at the end of this section. We notice that the Arrhenius plot reported in Fig. 4.14 provides also an excellent fit, with an activation energy of 2.5 ± 0.5 eV.

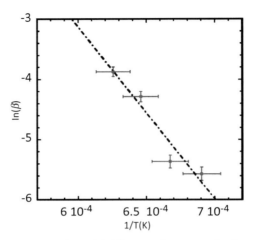

Figure 4.14 Arrhenius plot of ln(β) as a function of $1/T$, providing the activation energy of the SiC to graphene transformation [35].

To have an insight into this growth law and to understand the origin of the activation energy the authors developed a kinetic model, which takes into account the essential steps occurring during the epitaxial growth of graphene. To this end they referred to the bottom-up growth mechanism recently proposed in the literature, where graphene is continuously formed at the graphene-SiC interface during Si evaporation [22] (Fig. 4.12I). The chemical processes leading to the growth of the overlayer can be schematized as follows:

$$SiC \rightarrow (Si)_V + C^* \qquad (4.9)$$

$$C^* + G_n \rightarrow G_{n+1} \qquad (4.10)$$

Reaction 4.9 describes the evaporation of Si, which produces "reactive" carbon species at the interface, here denoted as C^*, and Si in the gas phase (Si_v). While Si diffuses through the holes of the layers above, leaving the solid, (Fig. 4.12) these reactive carbon units add to an already formed graphene layer (G_n) made up of n C units (Eq. 4.10).

The present approach considers the reactions above to be rate determining as far as the kinetics are concerned. Moreover the liberation of Si at the interface, for example, through the formation of vacancies and interstitials, is assumed to occur under steady-state conditions. The authors define rate equations for this model, linking the change in the number density of "free" C atoms n_{C^*} to the Si flux that leaves the interface J_{Si} and to the rate of graphene thickness increase $\dfrac{dh}{dt}$

$$\begin{cases} \dfrac{dn_{C^*}}{dt} = J_{Si} - K_2 n_{C^*} \\ \rho_G \dfrac{dh}{dt} = K_2 n_{C^*} \end{cases} \tag{4.11}$$

by means of the first-order rate constant of reaction (10) K_2, defining the attachment of C^* to the graphene network. The flux of Si atoms is modeled by means of Fick's first law as $J_{Si} = -D_{Si}\nabla_C$, which is approximated according to $J_{Si} = D_{Si}\dfrac{c_i}{h} = \dfrac{K_1}{h}$, where D_{Si} and $c_i \equiv c_{(Si)_i}$ are the diffusion coefficient of Silicon in graphene through defects and the concentration of mobile Si, respectively. By some manipulations [35] the system of equations (Eq. 4.11) becomes

$$\frac{1}{2}\frac{dz}{dh} + \sqrt{z} = \frac{B}{h}, \tag{4.12}$$

where

$$B = \frac{K_1}{K_2}\frac{1}{\rho_G h_0^2}, \tag{4.13}$$

$$\tau = K_2 t, \tag{4.14}$$

and

$$z(\tau) = \left(\frac{d\bar{h}}{d\tau}\right)^2. \tag{4.15}$$

The growth law is defined by the dimensionless growth speed $\sqrt{z}$ in terms of the dimensionless time τ, as the derivative of the dimensionless thickness $\bar{h}$. The constant B depends on the temperature through two rate constants, $K_1 \approx \exp\left[-\dfrac{U_d^* + \dfrac{E_{Si}}{2}}{k_B T}\right]$ and

$K_2 \approx \exp\left[-\dfrac{U_C^*}{k_B T}\right]$, where U_d^* is the activation energy for Si diffusion, E_{Si} is the energy for the liberation of Si atom, and U_C^* is the effective activation energy for the formation of graphene (Eq. 4.10). The behavior of z is found to depend strongly on B, namely on whether B is higher or lower than unity. The value $B = 1$ defines the transition between two growth regimes. For $B < 1$ the kinetics lead to the growth law

$$h(t) = \left(\frac{2K_1}{\varrho_G}\right)^{1/2} t^{1/2},\tag{4.16}$$

with the exponent equal to $1/2$. By combining Eqs. 4.4, 4.8, and 4.16, we find that $(\lambda \beta) = \left(\dfrac{2K_1}{\varrho_G}\right)^{1/2}$, and the Arrhenius plot (Fig. 4.14) provides an activation energy $E_a = \dfrac{U_d^* + \dfrac{E_{Si}}{2}}{2}$, which depends upon both kinetic and thermodynamic quantities.

In the growth regime where $B < 1$, the evaporation of Si (Eq. 4.9) is rate determining, whereas for $B > 1$, the transformation of C in graphene (Eq. 4.10) is rate determining. For $B > 1$ the kinetics exhibit a complex behavior that can be characterized, in a phenomenological way, by means of a time-dependent growth exponent.

There is also an intermediate range of growth, around $B \approx 1$, where both processes proceed at a comparable rate. On the basis of the XPS data analysis, which results in the time exponent $= 1/2$, the proposed approach lends support to a growth mechanism where the diffusion of Si through graphene is rate determining. In fact when $K_1 \ll K_2$ we have $B < 1$ (Eq. 4.13). This condition is well verified for the first three temperatures ($1175°C < T < 1275°C$) (see Fig. 4.13), where Eq. 4.5 and Eq. 4.6 provide an excellent fit to the experimental data with $n(t) = \beta \sqrt{t}$. At $T = 1325°C$ the fit worsens, suggesting that the condition $B < 1$ is not well satisfied in this case. At this temperature the growth law is expected to undergo a transition from $B < 1$ to $B \approx 1$, while the condition $B > 1$ is expected to hold at higher temperatures [25].

The authors note that the value E_a = 2.5 ± 0.5 eV calculated from the slope in Fig. 4.14 matches the result obtained in Ref. [31] and define this value as "effective" activation energy for the graphene growth. This number enters the rate constant K_1, for which they obtain the value $U_d^* + \dfrac{E_{Si}}{2}$, which is equal to $2E_a$ = 5 ± 1 eV. This value compares well within the experimental error, with the theoretical estimation of first principles calculations [21] (U_d^* = 4.7 eV and E_{Si} = 3.2 eV) providing $U_d^* + \dfrac{E_{Si}}{2} \approx 6 \pm 1 \text{eV}$.

4.3.5 Kinetics of Graphene Islands with Constant Thickness

It is interesting to compare the above results [35] with a model recently developed in the literature for graphene growth by Si sublimation. Norimatsu and coworkers [38] report that on C-face 6H-SiC(000$\bar{1}$) surface, graphene formation may occur through nucleation and lateral growth of graphene nuclei with a constant thickness. Specifically, sublimation of Si atoms occurs at preferential sites, leading to the formation of small craters on the surface. In fact, at surface defects such as atomic steps and screw dislocations, Si evaporation is expected to be easier. A graphene nucleus, three atomic layers in thickness, is typically formed within the crater independently of the temperature. After nucleation, Si atoms at the edge of the crater evaporate and the nucleus begins to grow, laterally. This process continues until the substrate surface is completely covered by graphene. After this stage the growth mechanism is expected to change, that is, the decomposition occurs along the normal direction. It is worth stressing that this study refers to temperature deposition above 1350°C and under high Ar pressure. In fact, the fraction of substrate surface covered by graphene at the end of deposition (30 min) is different from zero for $T > 1350°C$ and becomes equal to unity only at T = 1600°C.

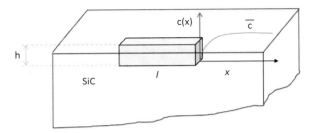

Figure 4.15 Growth model for graphene from a square nucleus of fixed thickness h and side l, embedded in the SiC substrate. The concentration profile, linked to C atom diffusion in the substrate, is also reported, where $\bar{c}$ is the C concentration far from the island border.

A pictorial view of this growth mode is reported in Fig. 4.15, which shows a nucleus, parallelepiped in shape with a square base of side l, embedded in the SiC matrix. In terms of $I_{ov}(t)$ and $I_{sub}(t)$ intensities (Eq. 4.3) such a growth mode implies a fixed attenuation factor $(1 - e^{-h/\lambda})$ for each island, multiplied by a surface coverage factor $S(t)$:

$$\frac{I_{ov}(t)}{I_{ov}^{\infty}} = S(t)(1 - e^{-h/\lambda}) \tag{4.17}$$

and

$$\frac{I_{sub}(t)}{I_{sub}^{0}} = [1 - S(t)(1 - e^{-h/\lambda})], \tag{4.18}$$

where $S(t)$ is the surface fraction covered by graphene and h the constant value of the thickness of graphene nuclei. In the expression above, I_{ov}^{∞} is the normalization factor (temperature dependent) that is equal to the asymptotic value of the intensity attained in the limit $h \gg \lambda$ and $S(t) = 1$. The term I_{ov}^{∞} is given by $I_{ov}^{\infty} = I\sigma\rho_s / \left(1 - e^{-\frac{a}{\lambda}}\right)$, where I is the incident photon flux, σ the photoemission cross section, ρ_s the surface density of C atoms in the graphene layer, and a the vertical lattice spacing. As mentioned above, I_{sub}^{0} is experimentally accessible since it is equal to $I_{sub}(0)$. On the other hand, the experimental determination of I_{ov}^{∞} is not straightforward, being constrained to the possibility of achieving the asymptotic conditions—$h \gg \lambda$ and $S(t) = 1$—during the growth. Nevertheless, this normalization term can be obtained by an appropriate analysis of the $I_{ov}(t)$ and $\dfrac{I_{sub}(t)}{I_{sub}^0}$ experimental "kinetics." In fact, combining Eq. 4.17 and Eq. 4.18 one obtains

$$I_{ov}(t) = I_{ov}^{\infty} \left[1 - \frac{I_{sub}(t)}{I_{sub}^0} \right], \qquad (4.19)$$

from which the normalization constant can be determined (at each temperature) by the slope of the straight line $I_{ov}(t)$ versus $\dfrac{I_{sub}(t)}{I_{sub}^0}$. Noteworthy is that Eq. 4.19 coincides with Eq. 4.7; in other words, it is independent of the growth mode considered. Therefore, the question arises on the possibility to have an insight into the growth mode through an XPS study.

A comprehensive and manageable approach to the $S(t)$ kinetics, which is able to deal with any surface coverage value, is represented by the Kolmogorov–Johnson–Mehl–Avrami (KJMA) model [39]. Accordingly, the kinetics are given by the stretched exponential $S(t) = 1 - e^{-Kt^m}$, where m gives the time dependence of the lateral growth law of the cluster. In the case of circular nuclei $r(t) = At^{m/2}$, r being the cluster radius. Incidentally, a stretched exponential function well describes the XPS data as discussed in the previous section.

It is worth recalling that the KJMA approach is based on the Poisson process and requires cluster nucleation to be random throughout the surface. In the model, nucleation and growth laws are assigned a priori; the first gives the rate at which stable clusters appear on the surface, the second the time dependence of the cluster size. The importance of the model relies on the possibility to describe the behavior of the fractional coverage up to the film closure when collision among clusters is important. Therefore, according to Poisson statistics, the probability that a point of the surface is not covered by the growing phase, at running time t, reads $P(t) = e^{-S_e(t)}$ where $S_e(t)$ is the convolution product between the nucleation rate $\dot{N}(t')$ and the cluster growth law. For a circular shape of the nuclei the growth law is expressed as $r(t - t')$, hich is the radius of the nuclei (that start growing at a time $t' < t$) at the actual time t. The $S_e(t)$ function is

$$S_e(t) = \pi \int_0^t \dot{N}(t')r^2(t-t')dt' \tag{4.20}$$

and is equal to the total area of the nuclei in the case of unimpeded growth [39]. In the case of a very fast nucleation process $\dot{N}(t') = N_0\delta(t')$, with $\delta(t')$ the Dirac delta and N_0 the surface density of nuclei, the $P(t)$ probability becomes $P(t) = e^{-\pi N_0 r^2(t)}$. Since the fractional coverage is just equal to $S(t) = 1 - P(t)$, for a power growth law the stretched exponential $S(t) = 1 - e^{-Kt^m}$ is obtained. It is worth recalling that the KJMA model also applies to a noncircular nucleus shape [39].

Equation 4.17 is used to describe the $I_{ov}(t)$ curves by employing the function $S(t) = 1 - e^{-Kt}$, which implies diffusional growth by C atoms to be rate determining ($S_e(t) \propto t^m$, $m = 1$). In fact, under these circumstances it is possible to show that the growth law is $l(t) \propto \bar{c}(D_C t)^{1/2}$, with $\bar{c}$ carbon concentration far from the island border and D_C the diffusion coefficient of C atoms[1] (see also Fig. 4.15). Both $(1 - e^{-h/\lambda})$ and K quantities have been determined by fit. The results of this analysis are reported in Fig. 4.16a, where, for $\lambda = 28$ Å [35] the constant thickness values are found in the range of 1–4 graphene layers [26]. Although the quality of the fit is quite good, the thickness is found to be at least a factor of two lower than the expected value [26]. Moreover, at odds with the layer-by-layer model, the kinetics Eq. 4.18, with the fitting parameters obtained from Fig. 4.16a, does not reproduce well the curves of the substrate spectra (Fig. 4.16b).

The rate constant K is linked to the whole activation energy of the process, E_a, according to $E_a \approx U_N + U_C + E_{Si}$, where U_N, U_C and E_{Si} are the activation energies for nucleation, C diffusion, and Si sublimation, respectively. The Arrhenius plot of the rate constants obtained by the fit of Fig. 4.16a provides $E_a = 2.2$ eV. This finding seems to underestimate, markedly, the actual value, since $E_{Si} \approx 3.5$ eV

---

[1]For linear diffusion, the solution of Fick's second law with initial and boundary conditions $c(x,0) = \bar{c}$ and $c(\infty, t) = \bar{c}$, $c(0, t) = 0$, respectively, leads to the growth flux $J = D\dfrac{\partial c}{\partial x}\Big|_{x=0} = \dfrac{\bar{c}\sqrt{D}}{\sqrt{\pi t}}$. Also, $\dfrac{dl}{dt} = \dfrac{J}{\rho}$, with ρ is the graphene density. The integration of this last equation gives $(t) \propto \bar{c}(D_C t)^{1/2}$.

[21, 35], and the activation energy for C diffusion in defective SiC is also expected to be quite high, being $U_C \approx 7$ eV for a SiC single crystal [40].

In the framework of the nucleation model discussed in Ref. [38], let us briefly digress on the growth law of graphene islands when Si evaporation is rate determining. As Si sublimation takes place at the boundary of the crater, the rate of Si evaporation $\frac{dn_{Si}}{dt}$ is proportional to the perimeter of the graphene phase $p\left(\frac{dn_{Si}}{dt} \propto kp\right)$, with k the rate constant. Since Si sublimation is rate determining, the growth law is attained by equating $\frac{dn_{Si}}{dt}$ to the rate of graphene growth: $\frac{dn_{Si}}{dt} = \frac{dn_C}{dt} = kp$.

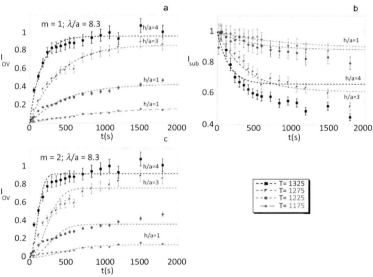

Figure 4.16 Fitting of the evolution of the overlayer intensity (a) and of the substrate intensity (b) as a function of annealing time following the island growth model proposed by Norimatsu (Eq. 4.17 and Eq. 4.18) for diffusion-type growth ($m = 1$ in the stretched exponential that describes $S[t]$). In (c) a similar analysis of the kinetics is displayed in the case of linear growth ($m = 2$ in the stretched exponential that describes $S[t]$).

Also, by denoting with $\dfrac{dl}{dt}$ the rate of lateral growth of the crater we get $\dfrac{dn_C}{dt} = \rho h p \dfrac{1}{2}\dfrac{dl}{dt}$, and the growth rate becomes $\dfrac{dl}{dt} = \text{const}$: the lateral growth of the crater is linear with time: $l \propto kt$. Accordingly, the fractional coverage could be given by the KJMA equation with $m = 2$. Under these circumstances, however, the quality of the fit worsens (Fig. 4.16c). In this case the overall activation energy of the process, which enters the rate constant K, is estimated as $E_a \approx U_N + 2E_{Si}$, to be compared with a value of $E_a = 3.4$ eV obtained by the Arrhenius plot of fitting parameters. Also in this case the result provides an underestimation of the expected value.

4.3.6 Alternative Kinetic Models

As evidenced in the previous section, a diffusion-type growth law of the form $h \propto \sqrt{t}$ is expected whenever the rate determining step of the reaction implies transport of matter across the growing phase. Also, nucleus growth can take place by surface diffusion of atoms toward the new phase. Such a mechanism could occur in the case of growth ruled by surface diffusion of C atoms at the edge of the graphene nucleus. In the following, we discuss two models for graphene island growth by C atom diffusion.

4.3.6.1 Terrace growth model

Let's assume C atom diffusion is rate determining and Si sublimation is fast. As a first example, we consider a terrace-shaped island whose edge grows by surface diffusion of C atoms (Fig. 4.17a). In this case we deal with a linear diffusion problem.

By denoting with $n_{s,c}$ the surface density of C units, the flux of C at the edge of the island is equal to $J = -D_C\left(\dfrac{\partial n_{s,c}}{\partial x}\right)_{x=0}$, where D_C is the surface diffusion coefficient of C and the graphene edge is located at $x = 0$. By denoting with L the length of the advancing front of the graphene phase, it follows that the rate of C atom incorporation into graphene terrace is $\dfrac{dn_C}{dt} = -JL$. The growth rate is eventually attained from the equation $\rho h L \dfrac{dX}{dt} = -JL$ that is $\dfrac{dX}{dt} = \dfrac{D_C}{h\rho}\left(\dfrac{\partial n_{s,c}}{\partial x}\right)_{x=0}$,

where h and ρ are the thickness and density of the graphene phase, respectively, and $\dfrac{dX}{dt}$ the rate of lateral growth. A quite simple solution of the present problem can be obtained by considering diffusion under steady-state conditions, that is, $\dfrac{\partial n_{s,c}}{\partial t} = 0$, for which the diffusion equation gives

$$\frac{\partial n_{s,c}}{\partial t} = D_C \frac{\partial^2 n_{s,c}}{\partial x^2} + I_C \approx 0 . \tag{4.21}$$

In Eq. 4.21 the term I_C is the rate, per unit area, of C generation at the surface linked to the reaction of Si sublimation. Equation 4.21 is solved with boundary condition $n(0) = n(b)$, where b is the distance between the edges of two graphene nuclei, supposed to be parallel (Fig. 4.17a). The solution of Eq. 4.21 provides a parabolic surface density profile, according to

$$n_{s,c}(x) = \frac{bI_C}{2D_C} x - \frac{I_C}{2D_C} x^2 . \tag{4.22}$$

The lateral growth rate, therefore, reads

$$\frac{dX}{dt} = \frac{bI_C}{2h\rho} , \tag{4.23}$$

which implies, for constant, b, a linear growth law. Moreover, for a growth rate lower than the characteristic diffusion time, the steady-state condition mentioned above can be assumed to hold at any time during the growth and the time dependence of b taken into account by means of the relation $2\dfrac{dX}{dt} = -\dfrac{db}{dt}$. Eq 4.23, therefore, provides

$\ddot{X} = -\dot{X}\dfrac{I_C}{h\rho}$, with solution $X(t) = \dfrac{b_0}{2}(1 - e^{-t/\tau}) + X(0)$, where $\tau = \dfrac{h\rho}{I_C}$

and $b_0 = b(0)$. Clearly, for τ values much longer than the duration of the growth the evolution of the terrace edge is linear in time. In addition, provided that steady-state conditions are reached, the kinetics of growth are independent of the surface diffusion coefficient D_C.

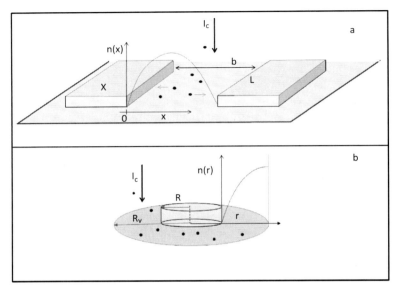

Figure 4.17 Two models for graphene growth: (a) terrace growth and (b) disk growth within a Voronoi cell. n(x) and n(r) denote the surface densities of C atoms as a function of the distance from the island border. At the steady state, the surface density profile is either parabolic (model a) or logarithmic (model b). I_c stands for the flux of C atoms either generated at the surface by thermal decomposition of SiC or incoming on the surface in a CVD process.

4.3.6.2 Disk growth model

A second case is related to the diffusional growth of disk-shaped graphene islands, in the framework of the capture zone model [41]. Each island belongs to a cell (Voronoi cell) that is defined as the space region whose points are closer to the considered island than to the others. In this model it is assumed that only C atoms belonging to the same cell of an island give rise to its growth (Fig. 4.17b). If steady-state conditions are established during the growth process, the following relation holds:

$$\rho h \pi \frac{dR^2}{dt} = I_C[a - \pi R^2], \qquad (4.24)$$

where I_C is the rate of C atom generation and R is the radius of the island within the Voronoi cell of area a. In Eq. 4.24 the first member represents the rate of island growth in C atoms per unit time and the second member gives the rate of C atom generation in the uncovered portion of the cell surface. The solution of the differential equation is $R^2 = \frac{a}{\pi}(1 - e^{-t/\tau})$, where $\tau = \frac{h\rho}{I_C}$. For a τ value much longer than the duration of the growth the parabolic growth law attained is $R = \left(\frac{aI_C}{\pi h\rho}\right)^{1/2} t^{1/2}$. Also, for such morphology of the film the kinetics of growth are independent of diffusion coefficient. On the other hand, the concentration profile of the C atom within the cell, at steady state, is a function of diffusion coefficient. This can be shown by solving the nonhomogeneous diffusion equation, at steady state, with the inclusion of the source term for C atoms: $D_C\nabla^2 n_{s,c} + I_C = 0$. Employing cylindrical coordinates the equation becomes

$$D_C \frac{1}{r}\frac{\partial(rn_{s,c})}{\partial r} + I_C = 0, \text{ that is}$$

$$\frac{1}{r}\frac{dn_{s,c}}{dr} + \frac{d^2 n_{s,c}}{dr^2} = -\frac{I_C}{D_C}, \tag{4.25}$$

where r is the distance from the center of the island. Equation 4.25 is solved for a circular Voronoi cell of radius R_V and with boundary conditions $n_{s,c}(R) = 0$, $\left.\frac{dn_{s,c}}{dr}\right|_{r=R_V} = 0$, R being the radius of the island (Fig. 4.17b). The surface density profile in the cell is eventually given by

$$n_{s,c}(r) = \frac{I_C}{2D_C}\left[R_V^2 \ln\frac{r}{R} - \frac{1}{2}(r^2 - R^2)\right], \tag{4.26}$$

which exhibits a logarithmic behavior. Notably, computation of the growth rate of the nucleus through the density profile, that is,

$$\frac{dR}{dt} = \frac{D_C}{\rho h}\left(\frac{\partial n_{s,c}}{\partial r}\right)_{r=R}$$, leads to the growth law previously obtained by exploiting mass conservation.

4.4 Conclusion

In summary, we have discussed here some possible growth modes of graphene on SiC substrates under Ar pressure and in UHV. The first two approaches, layer by layer and lateral extension of graphene islands with constant thickness, have been applied to describe experimental kinetics extracted from XPS data of thermal annealing of SiC. Both models point to a diffusional-growth law, where the characteristic size (either thickness or island radius) scales with time as $t^{1/2}$. Among the two models the first one seems to be more appropriate since (a) it provides more reasonable values of the derived physical quantities and (b) it exhibits a higher level of self-consistency.

The other two theoretical approaches presented deal with the growth of graphene terraces and islands on the surface of SiC by surface diffusion of C atoms under steady-state conditions. They are mainly appropriate for modeling graphene growth by CVD technique, where C atoms are supplied to the surface by the gas phase [34]. The present analysis shows that in both cases the growth law is exponential and reduces to linear or diffusion-type laws in the limit $t_{exp} << \tau$, where t_{exp} is the duration of the deposition.

This study provides a foundation to understand the kinetics of graphene growth on SiC, necessary to control the number and quality of graphene layers for technological applications.

Acknowledgments

The authors acknowledge the support of the Australian Research Council (ARC) through the Discovery Project DP130102120 and of the Australian National Fabrication Facility. The technical support of Dr. J. Lipton-Duffin and Dr. P. Hines of the Institute of Future Environments at QUT and the equipment of the Central Analytical Research Facility (CARF), as well as the funding by the ARC through the LIEF grant LE100100146, are also kindly acknowledged.

References

1. Emtsev, K. V., Bostwick, A., Horn, K., Jobst, J., Kellogg, G. L., Ley, L., McChesney, J. L., Ohta, T., Reshanov, S. A., Rohrl, J., Rotenberg, E., Schmid, A. K., Waldmann, D., Weber, H. B., and Seyller, T., Towards wafer-size graphene layers by atmospheric pressure graphitization of silicon carbide, *Nat. Mater.*, **8**(3), 203–207 (2009).

2. de Heer, W. A., Berger, C., Ruan, M., Sprinkle, M., Li, X., Hu, Y., Zhang, B., Hankinson, J., and Conrad, E. H., Large area and structured epitaxial graphene produced by confinement controlled sublimation of silicon carbide, *Proc. Natl. Acad. Sci. U S A*, **108**(41), 16901 (2011).

3. Ouerghi, A., Silly, M. G., Marangolo, M., Mathieu, C., Eddrief, M., Picher, M., Sirotti, F., El Moussaoui, S., and Belkhou, R., Large-area and high-quality epitaxial graphene on off-axis SiC wafers, *ACS Nano*, **6**(7), 6075–6082 (2012).

4. Tromp, R. M., and Hannon, J. B., Thermodynamics and kinetics of graphene growth on SiC(0001), *Phys. Rev. Lett.*, **102**(10), 106104 (2009).

5. Badami, D. V., Graphitization of alpha-silicon carbide, *Nature*, **193**(4815), 569–570 (1962).

6. Berger, C., Song, Z., Li, T., Li, X., Ogbazghi, A. Y., Feng, R., Dai, Z., Marchenkov, A. N., Conrad, E. H., First, P. N., and de Heer, W. A., Ultrathin epitaxial graphite: 2D electron gas properties and a route toward graphene-based nanoelectronics, *J. Phys. Chem. B*, **108**(52), 19912–19916 (2004).

7. Riedl, C., Starke, U., Bernhardt, J., Franke, M., and Heinz, K., Structural properties of the graphene-SiC (0001) interface as a key for the preparation of homogeneous large-terrace graphene surfaces, *Phys. Rev. B*, **76**(24), 245406 (2007).

8. Emtsev, K. V., Speck, F., Seyller, T., Ley, L., and Riley, J. D., Interaction, growth, and ordering of epitaxial graphene on SiC {0001} surfaces: a comparative photoelectron spectroscopy study, *Phys. Rev. B*, **77**(15), 155303 (2008).

9. Hass, J., Varchon, F., Millán-Otoya, J. E., Sprinkle, M., Sharma, N., de Heer, W. A., Berger, C., First, P. N., Magaud, L., and Conrad, E. H., Why multilayer graphene on 4HSiC(0001) behaves like a single sheet of graphene, *Phys. Rev. Lett.*, **100**(12), 125504 (2008).

10. Riedl, C., Coletti, C., Iwasaki, T., Zakharov, A. A., and Starke, U., Quasi-free-standing epitaxial graphene on SiC obtained by hydrogen intercalation, *Phys. Rev. Lett.*, **103**(24), 246804 (2009).

11. Ouerghi, A., Kahouli, A., Lucot, D., Portail, M., Travers, L., Gierak, J., Penuelas, J., Jegou, P., Shukla, A., and Chassagne, T., Epitaxial graphene on cubic SiC(111)/Si(111) substrate, *Appl. Phys. Lett.*, **96**(19), 191910-191910-3 (2010).

12. Riedl, C., Coletti, C., and Starke, U., Structural and electronic properties of epitaxial graphene on SiC (0 0 0 1): a review of growth,

characterization, transfer doping and hydrogen intercalation, *J. Phys. D: Appl. Phys.*, **43**, 374009 (2010).

13. Qi, Y., Rhim, S. H., Sun, G. F., Weinert, M., and Li, L., Epitaxial graphene on SiC(0001): more than just honeycombs, *Phys. Rev. Lett.*, **105**(8), 085502 (2010).

14. Goler, S., Coletti, C., Piazza, V., Pingue, P., Colangelo, F., Pellegrini, V., Emtsev, K. V., Forti, S., Starke, U., and Beltram, F., Revealing the atomic structure of the buffer layer between SiC (0001) and epitaxial graphene, *Carbon*, **51**, 249–254 (2013).

15. Nemec, L., Blum, V., Rinke, P., and Scheffler, M., Thermodynamic equilibrium conditions of graphene films on SiC, *Phys. Rev. Lett.*, **111**(6), 065502 (2013).

16. Emery, J. D., Detlefs, B., Karmel, H. J., Nyakiti, L. O., Gaskill, D. K., Hersam, M. C., Zegenhagen, J., and Bedzyk, M. J., Chemically resolved interface structure of epitaxial graphene on SiC(0001), *Phys. Rev. Lett.*, **111**(21), 215501 (2013).

17. Yazdi, G. R., Vasiliauskas, R., Iakimov, T., Zakharov, A., Syväjärvi, M., and Yakimova, R., Growth of large area monolayer graphene on 3C-SiC and a comparison with other SiC polytypes, *Carbon*, **57**, 477–484 (2013).

18. Mallet, P., Varchon, F., Naud, C., Magaud, L., Berger, C., and Veuillen, J. Y., Electron states of mono- and bilayer graphene on SiC probed by scanning-tunneling microscopy, *Phys. Rev. B*, **76**(4), 041403 (2007).

19. Wang, D., Liu, L., Chen, W., Chen, X., Huang, H., He, J., Feng, Y.-P., Wee, A. T. S., and Shen, D. Z., Optimized growth of graphene on SiC: from the dynamic flip mechanism, *Nanoscale*, **7**(10), 4522–4528 (2015).

20. Sforzini, J., Nemec, L., Denig, T., Stadtmüller, B., Lee, T. L., Kumpf, C., Soubatch, S., Starke, U., Rinke, P., Blum, V., Bocquet, F. C., and Tautz, F. S., Approaching truly freestanding graphene: the structure of hydrogen-intercalated graphene on 6H-SiC(0001), *Phys. Rev. Lett.*, **114**(10), 106804 (2015).

21. Sun, G. F., Liu, Y., Rhim, S. H., Jia, J. F., Xue, Q. K., Weinert, M., and Li, L., Si diffusion path for pit-free graphene growth on SiC (0001), *Phys. Rev. B*, **84**(19), 195455 (2011).

22. Huang, H., Chen, W., Chen, S., and Wee, A. T. S., Bottom-up growth of epitaxial graphene on 6H-SiC (0001), *ACS Nano*, **2**(12), 2513–2518 (2008).

23. Hannon, J. B., Copel, M., and Tromp, R. M., Direct measurement of the growth mode of graphene on SiC(0001) and SiC(000-1), *Phys. Rev. Lett.*, **107**(16), 166101 (2011).

24. Tanaka, S., Morita, K., and Hibino, H., Anisotropic layer-by-layer growth of graphene on vicinal SiC (0001) surfaces, *Phys. Rev. B*, **81**(4), 041406 (2010).

25. Drabińska, A., Grodecki, K., Strupiński, W., Bożek, R., Korona, K. P., Wysmołek, A., Stępniewski, R., and Baranowski, J. M., Growth kinetics of epitaxial graphene on SiC substrates, *Phys. Rev. B*, **81**(24), 245410 (2010).

26. Gupta, B., Notarianni, M., Mishra, N., Shafiei, M., Iacopi, F., and Motta, N., Evolution of epitaxial graphene layers on 3C SiC/Si (111) as a function of annealing temperature in UHV, *Carbon*, **68**, 563–572 (2014).

27. Li, L., and Tsong, I., Atomic structures of 6H SiC (0001) and (000$\bar{1}$) surfaces, *Surf. Sci.*, **351**(1), 141–148 (1996).

28. Emtsev, K., Speck, F., Seyller, T., Ley, L., and Riley, J., Interaction, growth, and ordering of epitaxial graphene on SiC {0001} surfaces: a comparative photoelectron spectroscopy study, *Phys. Rev. B*, **77**(15), 155303 (2008).

29. Rollings, E., Gweon, G. H., Zhou, S., Mun, B., McChesney, J., Hussain, B., Fedorov, A., First, P., de Heer, W. A., and Lanzara, A., Synthesis and characterization of atomically thin graphite films on a silicon carbide substrate, *J. Phys. Chem. Solids*, **67**(9), 2172–2177 (2006).

30. Tougaard, S., QUASES-IMFP-TPP2M program (http://www.quases.com/) based on S. Tanuma, CJ Powell and DR Penn, *Surf. Interf. Anal.*, **21**, 165–176 (1994).

31. Gupta, B., Notarianni, M., Mishra, N., Shafiei, M., Iacopi, F., and Motta, N., Corrigendum to "Evolution of epitaxial graphene layers on 3C SiC/Si (1 1 1) as a function of annealing temperature in UHV" carbon 68 (2014) 563–572, *Carbon*, **84**, 280 (2015).

32. Abe, S., Handa, H., Takahashi, R., Imaizumi, K., Fukidome, H., and Suemitsu, M., Surface chemistry involved in epitaxy of graphene on 3C-SiC (111)/Si (111), *Nanoscale Res. Lett.*, **5**(12), 1888–1891 (2010).

33. Fukidome, H., Abe, S., Takahashi, R., Imaizumi, K., Inomata, S., Handa, H., Saito, E., Enta, Y., Yoshigoe, A., and Teraoka, Y., Controls over structural and electronic properties of epitaxial graphene on silicon using surface termination of 3C-SiC(111)/Si, *Appl. Phys. Express*, **4**(11), 115104 (2011).

34. Pham, T. T., Santos, C. N., Joucken, F., Hackens, B., Raskin, J.-P., and Sporken, R., The role of SiC as a diffusion barrier in the formation of graphene on Si(111), *Diamond Relat. Mater.*, **66**, 141–148 (2016).

35. Zarotti, F., Gupta, B., Iacopi, F., Sgarlata, A., Tomellini, M., and Motta, N., Time evolution of graphene growth on SiC as a function of annealing temperature, *Carbon*, **98**, 307–312 (2016).

36. Sutter, P., Epitaxial graphene: how silicon leaves the scene, *Nat. Mater.*, **8**(3), 171–172 (2009).

37. Starke, U., Coletti, C., Emtsev, K., Zakharov, A. A., Ouisse, T., and Chaussende, D., Large area quasi-free standing monolayer graphene on 3C-SiC (111), *Mater. Sci. Forum*, **717–720**, 617–620 (2012).

38. Norimatsu, W., Takada, J., and Kusunoki, M., Formation mechanism of graphene layers on SiC (000$\overline{1}$) in a high-pressure argon atmosphere, *Phys. Rev. B*, **84**(3), 035424 (2011).

39. Fanfoni, M., and Tomellini, M., The Johnson-Mehl-Avrami-Kohnogorov model: a brief review, *Il Nuovo Cimento D*, **20**(7), 1171–1182 (1998).

40. Pelleg, J., *Diffusion in Ceramics* (series *Solid Mechanics and Its Applications*), Springer International, p. 448 (2016).

41. Tomellini, M., and Fanfoni, M., Scaling, voronoi tessellation and KJMA: the distribution function in thin solid films, *Int. J. Nanosci.*, **09**(01n02), 1–18 (2010).

Chapter 5

Atomic Intercalation at the SiC–Graphene Interface

S. Forti,[a] U. Starke,[b] and C. Coletti[a]

[a]*Center for Nanotechnology Innovation @ NEST, Istituto Italiano di Tecnologia, 56127 Pisa, Italy*
[b]*Max-Planck-Institut für Festkörperforschung, Heisenbergstr. 1, D-70569 Stuttgart, Germany*
camilla.coletti@iit.it

The electronic, chemical, and structural properties of epitaxial graphene obtained on $SiC(0001)$ and $SiC(111)$ can be manipulated by adopting an effective and robust technique that has been first presented in 2009 [1] and has gained popularity ever since: atomic intercalation at the $SiC(0001)$ interface. Atomic intercalation was initially demonstrated by using hydrogen (H_2) on $SiC(0001)$ [1] but can also be operated using different atomic species, such as gold (Au), germanium (Ge), lithium (Li), sodium (Na), oxygen (O_2), silicon

---

Growing Graphene on Semiconductors
Edited by Nunzio Motta, Francesca Iacopi, and Camilla Coletti
Copyright © 2017 Pan Stanford Publishing Pte. Ltd.
ISBN 978-981-4774-21-5 (Hardcover), 978-1-315-18615-3 (eBook)
www.panstanford.com

(Si), or fluorine (F) [2–9]. The resulting graphene layers—known as quasi-freestanding graphene—are in most instances electronically and structurally decoupled from the substrate. They can be neutrally charged or present n- or p-type doping. In this chapter we thoroughly discuss hydrogen intercalation and the properties of the resulting quasi-freestanding graphene. Specifically, Section 5.1 describes the structural and electronic properties of the interface layer; Section 5.2 discusses the basic principle of hydrogen intercalation and its practical implementation; in Sections 5.2.3, 5.2.4, and 5.2.5 the properties of quasi-free monolayer graphene (MLG), bilayer graphene (BLG), and trilayer graphene (TLG) on SiC(0001) are investigated; Section 5.2.6 presents hydrogen intercalation implemented on SiC(111) substrates; Section 5.2.7 discusses the impact of this techniques in science and technology; Section 5.3 examines other atomic intercalants that have been studied to date.

5.1 The Interface Layer

Before discussing hydrogen intercalation, it is necessary to describe the interface layer existing between SiC(0001) and graphene and to highlight its role. When a SiC(0001) surface is annealed at a high temperature, either under ultrahigh vacuum (UHV) or at higher pressure conditions, the first carbon layer developing—known as the buffer layer, interface layer, or zero-layer (ZL) graphene—presents a peculiar structure. The C atoms populating this layer represent a SiC(0001) carbon-rich reconstruction with a $(6\sqrt{3} \times 6\sqrt{3})R30°$ periodicity [10–12], and their positions geometrically correspond to the graphene honeycomb lattice. In Fig. 5.1, panels (a–c) show the unit cells of SiC(0001), graphene, and the superposition of the two, respectively. Panel (d) shows the supercell of the commensurate $(6\sqrt{3} \times 6\sqrt{3})R30°$ superlattice, which allows a graphene-like layer to arrange itself with a negligible lattice mismatch on top of SiC(0001). Figure 5.1e reports a scanning tunneling microscopy (STM) micrograph showing the honeycomb (graphene-like) structure of the interface layer. About 30% of the carbon atoms making up the interface layer are covalently bound to the Si atoms of the SiC substrate, as sketched in Fig. 5.2a. These bonds prevent the linear π-bands, a hallmark of graphene, to develop. Hence, this first carbon layer is electronically inactive. The formation of a second carbon layer proceeds by further decomposition of SiC (i.e., by heating SiC at higher temperatures). The second carbon layer emerges underneath the original interface layer and becomes itself the new interface layer [11]. Hence, the former interface layer becomes decoupled from the substrate. As a consequence, it now behaves as a true graphene monolayer and displays π-bands with the typical linear Dirac-like dispersion. This layer is indeed what is known as monolayer graphene on SiC(0001). The interface layer contains a high density of surface states that are at the origin of the strong n-type doping of as-grown monolayer graphene [10–13]. In addition, charged interface states are also believed to be responsible for the strongly reduced mobility in epitaxial graphene (EG) on SiC(0001) in comparison to exfoliated graphene flakes. Hence, for a practical application of EG on SiC(0001), it would be ideal to eliminate the interface bonding completely.

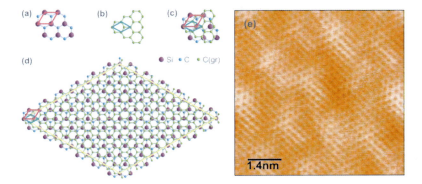

Figure 5.1 (a) Schematic model of the SiC(0001) unit cell (red diamond) in top view. (b) Schematic model of a graphene layer, where the unit cell is indicated by the green diamond. (c) Superposition of a SiC layer and a graphene layer with a 30° mutual rotation. The carbon atoms in SiC and in graphene [C(gr)] are presented in different colors to distinguish the two layers. (d) Commensurate superstructure of graphene on SiC(0001), where the $(6\sqrt{3}\times6\sqrt{3})R30°$ unit cell is indicated by the yellow diamond. (e) STM image of the interface layer imaged with a sample bias of −0.223 V (a–d) From Ref. [14], reproduced with permission. (e) Reprinted from Ref. [12], Copyright (2013), with permission from Elsevier.

5.2 Hydrogen Intercalation

5.2.1 How It Works

Elimination of the covalent bonding at the interface in order to decouple the EG layers from the SiC substrate requires the breaking and passivation of the respective bonds. As sketched for zero-layer graphene (ZLG) in Fig. 5.2, hydrogen intercalation can induce the desired decoupling. The Si atoms in the uppermost SiC bilayer can be saturated with atomic hydrogen (Fig. 5.2b). In this way the ZL is turned into a quasi-freestanding monolayer graphene (QFMLG). It should be noted that the process of hydrogen intercalation is not restricted to the initial ZL on SiC. Rather, the interface layer can be decoupled for mono- and bilayer graphene as well, yielding quasi-freestanding bilayer graphene (QFBLG) (see Section 5.2.4) and quasi-freestanding trilayer graphene (QFTLG) (see Section 5.2.5), respectively. In general, the transformation of n-layer graphene films into $(n + 1)$-layer graphene films is achieved [1]. Technically, hydrogen intercalation is a process that is carried out after graphene growth, as it is described in the next section.

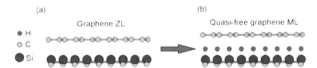

Figure 5.2 Side-view models of (a) as-grown zero-layer graphene on SiC(0001) and (b) quasi-freestanding monolayer graphene on SiC(0001) obtained after hydrogen intercalation of a zero-layer.

5.2.2 Technical Details

Before performing graphene growth and subsequent hydrogen intercalation nominally on-axis-oriented SiC(0001) surfaces need to be rendered atomically flat. This can be done by carrying out hydrogen etching at temperatures above 1300°C, as described in [15, 16]. The EG layers can then be prepared by SiC graphitization under UHV conditions [17] or in a furnace under atmospheric pressure (AP) [18]. Hydrogen intercalation is performed by annealing ZL graphene at temperatures between 600°C and 1000°C in molecular hydrogen at AP. While temperatures between 600°C and 800°C have been found to be sufficient to successfully intercalate ZLG, the use of higher temperatures (up to a maximum of 1000°C) is instrumental in achieving a complete intercalation when working with thicker graphene layers [19]. The use of temperatures higher than 1000°C might induce atomic defects in the graphene lattice. The intercalation process was originally carried out in a quartz-glass horizontal reactor in an atmosphere of palladium-purified ultrapure molecular hydrogen [1], similar to the technique used for hydrogen etching [16] and hydrogen passivation [20–22] of SiC surfaces. However, it should be noted that hydrogen intercalation is a robust process that can be successfully carried out within a variety of different reactors (hot wall or cold wall and horizontal or vertical). Therefore, the process parameters reported above can be successfully adapted to perform hydrogen intercalation also when using other systems.

5.2.3 Quasi-Freestanding Monolayer Graphene

As tight thickness control and large-scale domains are prerequisites for the successful fabrication of electronic devices, SiC(0001) is considered a suitable surface to obtain graphene for electronic applications. However, two hurdles might sensibly limit the prospects of graphene on SiC(0001). First, as a result of the graphene/SiC interface properties, as-grown EG is strongly electron doped. This doping translates, in terms of electronic band structure, into a measurable displacement (typically 0.4 eV) of the Fermi level, E_F, above the crossing point of the π-bands (Dirac point) [17, 23, 24]. Second, the mobilities typically measured for graphene on SiC(0001) are significantly lower than those found for exfoliated flakes and on SiC(000$\bar{1}$) substrates [25]. Hydrogen intercalation is an effective and elegant way of overcoming such drawbacks. When intercalating hydrogen at the interface, the intrinsic electron doping of EG can ultimately be overcome and the carrier mobilities might be sensibly increased.

To demonstrate the effect of the hydrogen treatment process, Fig. 5.3 shows angle-resolved photoemission spectroscopy (ARPES) measurements around the $\bar{K}$-point of the graphene Brillouin zone (BZ). For ZLG, no π-bands are observed (Fig. 5.3a). After hydrogen treatment the decoupling is clearly evident since the linear dispersing π-bands of monolayer graphene appear (Fig. 5.3b). Hence, it is clear that the hydrogen atoms migrate under the covalently bound initial carbon layer, break the bonds between C and Si, and saturate the Si atoms, as sketched in Fig. 5.2b. Consequently, after intercalation, the former ZL displays the electronic properties of QFMLG [1]. However, it should be noted that the decoupled graphene is typically hole-doped [26]. The exact doping level can vary somewhat and can reach up to 4×10^{12} cm^{-2}, corresponding to a shift of E_F of 300 meV below the Dirac point energy, E_D [27]. This variation can be attributed to possible instabilities of the intercalation process itself, such as the presence of isolated unsaturated silicon dangling bonds that counterdope the p carriers with n charge, and also to the presence of chemisorbed species from air. Indeed, we found that, starting from a fully intercalated sample, isolated hydrogen atoms could be partially desorbed upon annealing at ~700°C. As a consequence, presumably through this charge injection connected to single sites, the π-bands are gradually filled until charge neutrality ($E_F = E_D$) is reached (Fig. 5.3c). A possible explanation for the measurable p-type doping in freestanding graphene has been forwarded by Ristein et al. [28] as being the result of the spontaneous polarization of the hexagonal SiC polytypes.

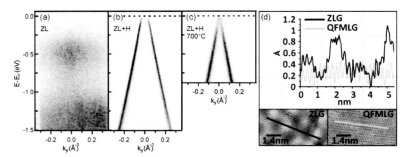

Figure 5.3 Dispersion of the π-bands measured with ARPES at the $\overline{K}$-point of the graphene Brillouin zone (a) for an as-grown graphene zero layer (ZL) on SiC(0001) and (b) after hydrogen treatment and subsequent annealing to (c) 700°C (Adapted (figure) with permission from Ref. [26]. Copyright (2011) by the American Physical Society). (d) Height profile extracted from the STM images shown at the bottom of the panel, corresponding to zero-layer and H-intercalated EG (Reprinted from Ref. [12], Copyright (2013), with permission from Elsevier).

Further evidence of the structural decoupling of quasi-freestanding graphene can be obtained via STM analysis. Figure 5.3d illustrates a roughness analysis of the graphene surface before and after hydrogen intercalation. A line profile analysis of the ZL image is shown by the black line in Fig. 5.3d and demonstrates the high corrugation of this interface layer, which is a result of the spatially varying coupling to the SiC substrate mediated by covalent bonds, as noted above. The peak-to-peak corrugation value of about 1 Å agrees with the theoretical value calculated by Varchon et al. [29]. As shown by the light-gray line in Fig. 5.3d, in the case of the QFMLG sample, the line profile analysis yields a peak-to-peak corrugation of approximately 0.4 Å, demonstrating that QFMLG is extremely flat. Indeed, for QFMLG the peak-to-peak corrugation is dominated by the graphene lattice (about 0.3 Å), while the residual long-range variations are around 0.1 Å [12].

It should be mentioned that hydrogen intercalation works for surfaces of different homogeneities. Indeed, although intercalation of hydrogen atoms might initiate at defective sites, highly homogeneous graphene synthesized at AP can be intercalated as well as UHV-grown graphene with poor homogeneity. The low-energy electron microscopy (LEEM) micrographs displayed in Fig. 5.4 show the intercalation process performed on UHV-prepared and AP-grown surfaces. Indeed, LEEM is a precious tool to analyze with spatial resolution the effect of hydrogen intercalation. In thin graphene slabs, the number of graphene layers in different areas can be determined by measuring LEEM reflectivity curves in the energy range of the graphitic conduction band (about 0–7 eV), with the number of dips corresponding to the number of layers [30]. A visible difference in grayscale contrast for the UHV-prepared sample (first two columns) is indicative of poor thickness homogeneity, as confirmed by the low-energy reflectivity curves. Differently, the sample prepared via AP is—as expected [18]—highly homogeneous in thickness, with a small inhomogeneity only at step edges. The subtle contrast difference for different terraces in the AP-graphene images stems from different stacking terminations of the 6H-SiC substrate (ABC versus CBA) underneath. The *IV* curves and the inset in the third image demonstrate that the number of graphene layers is the same in these areas. It is evident for both preparation methods that the areas presenting $n + 1$ layers, upon hydrogen intercalation, consistently become thick n-layers after hydrogen is desorbed by heating at high temperatures. The hydrogen intercalation process is indeed totally reversible. Si-H bonds are known to break at temperatures just above 700°C [31] and indeed, upon UHV annealing, complete deintercalation can be achieved [1]. Specifically, we found via STM analysis that when annealing at a temperature of about 750°C large surface areas deintercalated coherently and retransformed into patches of ZLG structure [14, 26]. At around 900°C, ARPES as well as X-ray photoelectron spectroscopy (XPS) measurements confirm that the ZL structure is completely re-established.

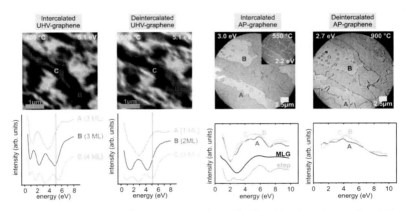

Figure 5.4 Summary of LEEM topography and reflectivity for H-intercalated EG in the cases of UHV and AP grown. The top panels label the underlying data. The energy, temperature, and scale at which the LEEM micrographs were recorded are highlighted on every panel. Furthermore, the intercalated AP-graphene LEEM panel hosts an inset of a LEEM micrograph taken on the same area at an energy in which the contrast coming from different areas is the same. In the bottom panels the LEEM *IV* reflectivity curves referred to each LEEM-imaged area are reported. The spot where the *IV* curves were recorded are labeled on every LEEM micrograph as A, B, and C (if any). Adapted (figures) with permission from Refs. [1, 26]. Copyright (2009, 2011) by the American Physical Society.

5.2.4 Quasi-Freestanding Bilayer Graphene: Seeking a Bandgap

As a consequence of the electric dipole existing at the graphene/SiC interface, BLG on SiC(0001) presents a bandgap of approximately 0.1 eV [17, 23], which is an additional appealing feature of EG on

SiC(0001) as it suggests the possibility of implementing field effect transistors (FETs) with an off state. Previous work has shown that by adopting molecular doping or by applying a perpendicular electric field this bandgap can be significantly enlarged [32–37]. Hydrogen treatment of a monolayer graphene yields an AB-stacked QFBLG, as demonstrated by the band structures displayed in Fig. 5.5a,b. Again, after intercalation the sample displays hole doping, estimated to be approximately 5×10^{12} cm^{-2}. Annealing at about 700°C significantly reduces the doping (see Fig. 5.5c). This indicates the vanishing of the electrostatic potential difference between the lower and upper layer in QFBLG. However, as for exfoliated graphene flakes, a bandgap can be engineered either by application of a perpendicular electric field, which breaks the inversion symmetry of graphene [32, 33, 36] or by deposition of a chemical dopant [34, 35, 37].

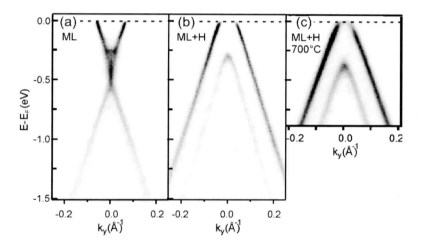

Figure 5.5 Dispersion of the π-bands measured with ARPES at the K̄-point of the graphene Brillouin zone for (a) an as-grown monolayer (MLG) and (b) after hydrogen treatment and annealing to (c) 700°C. The energy range in (c) is narrowed to better highlight the variation of doping level as a consequence of the annealing. (a) Adapted (figure) with permission from Ref. [26.] Copyright (2011) by the American Physical Society.

Molecular doping [38] is an effective way to control doping and bandgap magnitude in BLG [34, 37]. It has been demonstrated that upon functionalization with the strong molecular acceptor tetrafluorotetra-cyanoquinodimethane (F4-TCNQ) (molecule

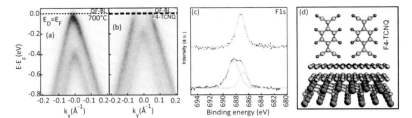

Figure 5.6 Dispersion of the π-bands measured with ARPES at the K̄ point of the graphene Brillouin zone for (a) a quasi-freestanding bilayer graphene on SiC(0001) obtained via hydrogen intercalation and annealed at 700°C, (b) the same bilayer after deposition of a multilayer of F4-TCNQ, and (c) F 1s core-level photoelectron emission spectrum for a multilayer of F4-TCNQ deposited on as-grown bilayer graphene (top spectrum) and quasi-freestanding bilayer graphene (bottom spectrum). The experimental data are displayed in black dots. The gray solid line is the envelope of the fitted components. (d) Schematic drawing of F4-TCNQ deposited on graphene. From Ref. [119] with permission of Springer.

sketched in Fig. 5.6d) the original bandgap present in BLG can be more than doubled [34]. Hence, it should be interesting to investigate whether deposition of F4-TCNQ molecules on hydrogen (H)-intercalated graphene leads to a doping effect comparable to that observed in as-grown graphene. To this end, an AP-grown monolayer graphene sample was first H-intercalated. After annealing at 700°C in UHV, charge neutrality was achieved, as shown by the ARPES spectrum reported in Fig. 5.6a. F4-TCNQ molecules were subsequently deposited from submonolayer to multilayer coverage, thus inducing a progressive upshift of the parabolic bands. The maximum upshift measured is displayed in Fig. 5.6b. Clearly, the molecules hole-dope the initially charge-neutral bilayer sample (the Dirac point is located about 200 meV above the Fermi level). It should be mentioned that when deposited on a QFBLG, F4-TCNQ molecules arrange in a different fashion than when deposited on as-grown BLG. This is clearly indicated by the XPS data in Fig. 5.6c, where the F 1s core-level spectra measured after deposition of a multilayer of F4-TCNQ on an as-grown bilayer (top trace) and a quasi-free bilayer (bottom trace) are reported. On an as-grown bilayer, the F 1s spectrum is symmetric, in agreement with what was reported in [34]. On the other hand, for QFBLG two components are clearly distinguishable, indicating that different F species exist in the molecular film and that, different from the case of as-grown bilayer [34], not only the cyano groups of the molecule are responsible for the charge transfer process. The different molecular arrangement might be induced by differences in terms of corrugation and long-range rippling between the as-grown and the QFBLGs. Although, ARPES analysis does not allow one to appreciate the band structure around the Dirac point, it is reasonable to assume that—thanks to the additional dipole developing at the interface upon molecular functionalization—a bandgap can be opened.

5.2.5 The ABC of Quasi-Freestanding Trilayer Graphene

In recent times, TLG has attracted wide attention owing to its stacking-and-electric-field-dependent electronic properties [39–46]. TLG has two naturally stable allotropes characterized by either Bernal (ABA) or rhombohedral (ABC) stacking of the individual carbon layers. In ABA stacking the atoms of the topmost layer obtain lateral positions exactly above those of the bottom layer (Fig. 5.7a). In an ABC-stacked trilayer each layer is laterally shifted with respect to the layer below by a third of the diagonal of the lattice unit cell (Fig. 5.7b). Several theoretical studies have predicted the electronic dispersion of ABA- and ABC-stacked trilayers using tight-binding approaches [39–41, 47–51]. The low-energy band structure of ABA TLG consists of a linearly dispersing (monolayer-like) band and bilayer-like quadratically dispersing bands (Fig. 5.7c) [39, 41, 49]. Quite differently, ABC trilayers have a single low-energy band with approximately cubic dispersion (Fig 7(d)) [39–41, 50]. A very intriguing distinction between the two allotropes is their behavior in the presence of a perpendicular electric field: ABA-stacked trilayers are expected to display a tunable band overlap, while ABC-stacked trilayers present a tunable bandgap, the latter being very appealing for electronic applications [41, 48]. However, the alluring rhombohedral phase is quite rare in nature as the energetically favored Bernal stacking makes up for more than 80% of the existing graphite [52, 53]. On the experimental side, progress in revealing the fundamental properties of TLG has been slow as such studies require homogeneous trilayers with a well-defined stacking sequence over areas of hundreds of micrometers.

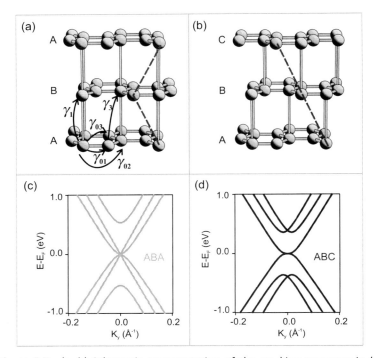

Figure 5.7 (a, b) Schematic representation of the stacking sequence in (a) Bernal and (b) rhombohedral TLG. The interatomic tight-binding hopping parameters between adjacent layers—thus valid for both stackings—are denoted by the black arrows in panel (a). (c, d) Calculated low-energy band structure around the K̄-point for Bernal (c) and rhombohedral (d) TLG. Reprinted (figure) with permission from Ref. [19.] Copyright (2013) by the American Physical Society.

Infrared conductivity and transport measurements have confirmed that a bandgap can be opened in ABC-stacked TLG when applying a perpendicular electric field, while no bandgap has been observed in ABA-stacked trilayers [43]. However, direct visualization of the electronic band structure of homogeneous TLG via ARPES was reported only in 2013 by our group [19]. In 2007, Ohta and colleagues reported ARPES spectra of UHV grown few-layer graphene on SiC [54], where the challenging separation of contributions from areas with a different number of layers or different stacking was presented. By performing hydrogen intercalation on highly homogeneous BLG we were able to obtain high-quality TLG samples, which have been instrumental for detailed band structure studies. Figure 5.8a displays a typical LEEM micrograph for as-grown BLG on 6H-SiC(0001). As discussed in Section 5.2.3, areas with differences in the reflected intensity identify regions with different graphene thicknesses (see Fig. 5.8c). Notably, the sample is highly homogeneous with, medium-gray domains (Fig. 5.8b) occupying more than 80% of the overall area. In Fig. 5.8d, XPS C 1s spectra measured before and after the hydrogen intercalation of a BLG are shown. The raw data appear in the figure as black dots, whereas the fitted components and their envelope are represented by solid curves. Notably, after the intercalation only the graphitic component remains, as it occurs for the ZLG and MLG [1, 26].

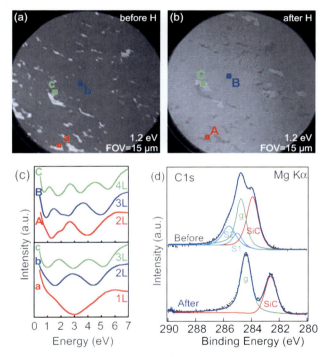

Figure 5.8 Representative LEEM micrographs with a field of view (FOV) of 15 μm recorded with an electron energy of 1.2 eV for (a) as-grown BLG on 6H-SiC(0001) and (b) the same area after hydrogen intercalation. (c) Electron reflectivity curves collected for the labeled regions (in panels a and b) of the initial surface (bottom graph) and of the H-intercalated graphene sample (top graph). (d) C 1s spectra of as-grown bilayer graphene on 6H-SiC(0001) before (top) and after hydrogen intercalation (bottom). The raw data is plotted as black dots and the envelope of the fitted components as a continuous line. Reprinted (figure) with permission from Ref. [19.] Copyright (2013) by the American Physical Society.

154 | *Atomic Intercalation at the SiC–Graphene Interface*

The band structure measured for this sample is shown in Fig. 5.9a. The spectrum is extremely sharp and exclusively consists of parabolic bands, the signature of BLG, corroborating the extreme homogeneity of the graphene film. In Fig. 5.9d, theoretical bands obtained by tight-binding calculations for a Bernal-stacked bilayer using the formalism of McCann and Fal'ko [55] are fitted to the experimental data. As expected for epitaxial BLG on SiC, the Fermi level is shifted by around 0.3 eV above the Dirac energy of the π-bands—indicative of n-type doping [23, 34]. Also, the characteristic bandgap of 120 meV caused by the electrostatic asymmetry of the bilayer slab on the SiC substrate is visible [23, 34]. The LEEM micrograph in Fig. 5.8b shows the same sample area as in Fig. 5.8a but upon annealing the sample in hydrogen. Indeed, the number of dips in the electron reflectivity spectra reported in the upper panel of Fig. 5.8c confirms the conversion of all the *n*-layers into (*n* + 1) layers.

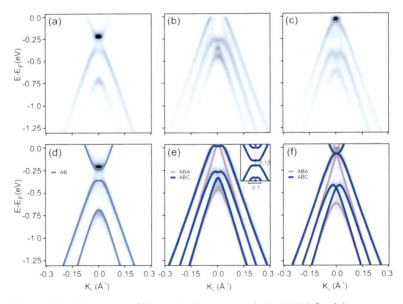

Figure 5.9 Dispersion of the π-bands measured via ARPES for (a) as-grown BLG on 6H-SiC(0001), (b) QFTLG annealed at 400°C, and (c) at about 800°C. The spectra are measured with a photon energy of 90 eV and with scans oriented perpendicular to the $\overline{\Gamma K}$-direction of the graphene Brillouin zone. (d–f) Tight-binding bands fitted to the experimental data shown in (a), (b), and (c), respectively. The fitting retrieves a bandgap in the ABC dispersion in (b) of 120 meV (the inset in [e]). Reprinted (figure) with permission from Ref. [19.] Copyright (2013) by the American Physical Society.

The first well-resolved direct visualization of the electronic band structure of TLG is displayed in Fig. 5.9b. As for QFMLG and QFBLG, the sample was annealed at a mild temperature of 400°C, which is sufficient to remove air contamination but well below the onset of hydrogen desorption [26]. Fitting of the experimental bands allows for a precise identification of the trilayer stacking sequence [19]. Figure 5.9e shows the results of the fitting procedure superimposed to the electronic dispersions obtained experimentally. The two stacking sequences, ABA and ABC, can be clearly distinguished as indicated by the respective magenta and blue fitting curves. The accurate overlap of the calculated bands with the experimental data reveals unambiguously that QFTLG on SiC contains domains of both Bernal and rhombohedral stacking, in contrast to natural graphite, which typically only features ABA stacking [52, 53]. The p-type doping typical of H-intercalated samples on α-SiC is observed (Section 5.2). This polarization induces an electrostatic field across the trilayer slab. Indeed, the fits indicate the presence of an energy bandgap with a magnitude of about 120 meV (the inset in [e]) [19]. This value is in agreement with results from infrared conductivity measurements presented elsewhere [43]. Similar to the cases reported in Sections 5.2.3 and 5.2.4; also in this case it is possible to achieve charge neutrality within a few meV by simply annealing the sample (although a slightly higher temperature is needed, that is, about 800°C). Indeed, the sample appears to have acquired a minimal n-doping after this treatment by possibly desorbing an excess of hydrogen from the Si dangling bonds. The visibility of the onset of the conduction band allows one to appreciate the absence of a measurable bandgap. Interatomic hopping parameters have been retrieved from a direct comparison of the experimental electronic and theoretical bands and are reported in [19]. Notably, a simple visual inspection suggests that the intensities of the ABC bands of all the measured spectra are higher than those of the ABA contributions. Detailed analysis of the intensities of the momentum distribution curves (MDCs) is presented in Ref. [19] and quantitatively confirms that the ABC branches are significantly more intense than the ABA ones. Of course, it must be taken into consideration that the photoemission intensity of single ABA and ABC branches is expected to vary as a consequence of varying strength and direction of interatomic interactions [54, 56]. Nevertheless, these results suggest that the ABC type of stacking occurs in QFTLG on SiC with a significantly higher incidence than in nature. The tendency of graphene to form on SiC in ABC stacking was experimentally observed by means of high-resolution transmission electron microscopy (HRTEM) measurements (see Fig. 5.10) and could be explained by a weakening of the $\gamma 5$ interatomic interaction—a major contributor to the stability of the ABA stacking—due to the displacement of carbon atoms in the buffer layer during the growth process [57].

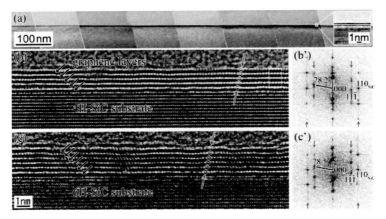

Figure 5.10 HRTEM images of graphene layers on (a) and (c) 6H- and (b) 4H-SiC. In the low-magnification image, together with the inserted high-magnification image in (a), the electron beam is incident parallel to the [11$\bar{2}$0] direction of SiC, while in (b) and (c), it is parallel to the [1$\bar{1}$00] direction of SiC. Graphene layers are observed as the dark lines indicated by the black arrows. A simulated HRTEM image of ABC stacking is shown in the inset in (b). In (b′) and (c′), FFT patterns of the area around (b) and (c) are shown. Reproduced (figure) with permission from Ref. [57]. Copyright (2010) by the American Physical Society.

As on SiC substrates the occurrence of the electronically appealing ABC stacking appears significantly higher than in natural bulk graphite, TLG on SiC might be the material of choice for the fabrication of a new class of gap-tunable devices.

5.2.6 Hydrogen Intercalation at the 3C-SiC(111)/Graphene Interface

With its extreme robustness and proven biocompatibility [58] cubic SiC (3C-SiC) is an appealing platform for the growth of graphene that could lead to a new generation of microelectromechanical systems and advanced biomedical devices. Moreover, cubic SiC can be epitaxially grown on Si crystals and—provided the process temperatures can be sufficiently lowered—this could reduce the production costs of graphene. The [111] orientation of this crystal naturally accommodates the sixfold symmetry of graphene. It has been shown that the structural and electronic properties of graphene obtained on 3C-SiC(111) are comparable to those of graphene on SiC(0001) [59, 60]. The first carbon layer that grows on top of SiC(111) is indeed a ZL as that developing on SiC(0001) surfaces. Thanks to advances in the heteroepitaxy of 3C-SiC(111) it is now possible to obtain thick, highly crystalline epilayer films [61] that allow the growth of high-quality graphene. It has been recently shown that hydrogen intercalation allows one to obtain highly homogeneous graphene with domains extending over areas of hundreds of square micrometers [62]. Notably, not only QFMLG has been demonstrated on such surfaces but also QFTLG [19]. It should be noted that when implementing hydrogen intercalation on cubic SiC, the resulting quasi-freestanding graphene is slightly n-type doped or neutrally charged [19, 62]. A likely explanation for this behavior is the one presented by Ristein et al.: the absence of the spontaneous polarization of the substrate imposed by hexagonal SiC's pyroelectricity [28]. Another peculiar characteristic of quasi-freestanding graphene on 3C-SiC(111) is that the band velocity retrieved from ARPES measurements is higher than that measured for graphene on hexagonal polytypes. As an example, the band velocity of the rhombohedral QFTLG on 6H-SiC(0001) is about 0.93 $\times$ 10^6 m/s, while on 3C-SiC(111) it is about 1.05 $\times$ 10^6 m/s [19]. It is still not clear at the moment whether these differences arise from a different concentration of scattering centers or are polytype dependent.

5.2.7 Hydrogen Intercalation: Impact and Advances

Hydrogen intercalation is a robust technique: it is easy to reproduce in different furnaces, and the resulting samples have excellent temperature stability and are not affected by exposure to air. Hence, since its first demonstration, this technique has been widely adopted both for fundamental and application-oriented studies. As previously discussed, the presence of charged interface states in the ZL is responsible for the reduced carrier mobility in as-grown graphene on SiC(0001). Also, attaining charge neutrality is vital to approaching the high carrier mobilities measured for freestanding or suspended graphene [63]. Indeed, removal of the interface layer yields a sensible increase of the carrier mobility. Transport measurements performed on QFMLG indicate an increase of the carrier mobility of at least 1 order of magnitude with respect to that measured for as-grown monolayer graphene (i.e., more than 11,000 $cm^2 V^{-1} s^{-1}$ at 0.3 K) [64]. In a different work, the hydrogen annealing temperature was found to significantly affect the measured carrier mobility. The highest mobilities were obtained because of reduced charged impurities when annealing between 700°C and 800°C [65]. Also, the use of hydrogen-intercalated samples has allowed the implementation of the first bottom-gated EG devices [66]. Intercalation has a profound influence on the detailed band structure parameters. Thus, for instance, one finds a renormalization of the π-band velocity of about 3% at energies approximately 200 meV below the Fermi level, which originates from electron–phonon interactions [67]. Simulation of the spectral function observed in the experiments allowed analysis of the coupling of the Dirac electrons to different quasi-particles in different energy regimes [26, 68]. Furthermore, through intentional n doping of hydrogen-intercalated QFMLG samples on SiC(0001), so-called plasmarons, representing the interaction of plasmons with electron–hole pairs, were observed by ARPES for the first time [27, 69]. Finally, it shall be noted that hydrogen intercalation is a fully scalable technique. The outstanding properties of graphene can be made accessible in quasi-freestanding EG layers on large-scale SiC(0001) wafers suitable for practical technological applications. Indeed, it was shown that fabrication of graphene transistors with a gain cutoff frequency of 24 GHz can be achieved on a wafer scale [70].

5.3 Intercalation of Different Atomic Species

Graphene can also be grown on many transition-metal surfaces; hence, the study of the interactions of transition metals with EG on SiC(0001) is of considerable interest. Indeed, gold was found to intercalate between ZLG and SiC(0001) [2]. In this case, two doping levels can be achieved; one phase has weak p doping, and the other exhibits strong n doping. This discovery triggered theoretical investigations that verified the experimental findings [71], more detailed measurements of π-band parameters [27], and further theoretical efforts to study transition metals [72]. Similar to gold, also atomically thin layers of germanium can be intercalated at the SiC/graphene interface, yielding two symmetrically doped (n- and p-type) phases, depending on the Ge coverage [3]. Such phases can be prepared individually or in coexistence. Hence, by intercalating Ge, lateral p/n junctions can be generated on EG with their size tailored on a mesoscopic scale [3, 73] and exotic effects such as Klein tunneling [74] can be observed [75].

Selected alkali metals have been found not only to serve as strong n-type transfer dopants but also to intercalate between graphene layers and the SiC substrate [4, 76]. Notably, Ref. [5] reports that lithium intercalates immediately upon room-temperature deposition, whereas moderate annealing to 75°C is necessary for sodium. Nevertheless, both lithium and sodium still induce strong n doping in the decoupled graphene layer [5], as also corroborated by theoretical work [77]. In contrast to Ref. [5], Sandin et al. reported that sodium intercalation takes place even at room temperature (accompanied by electron doping of the graphene) [6]. At low coverages and room temperature, they found that intercalation occurred through small chain-like structures and that the sodium was inserted between single-layer graphene and the interfacial layer. Sodium deposited at higher coverages and then annealed formed a second intercalation structure in which the sodium penetrated beneath the interfacial layer and decoupled it to form a second graphene layer. The two phases were observed to coexist [6]. Recently, silicon was also found to intercalate between the interface layer and the SiC substrate, but only upon annealing of the deposited silicon adlayer to about 800°C [7].

Finally, intercalation of highly electronegative atoms, such as N, O, and F, has been recently addressed. F intercalation has been reported to result in strong p-type doping [78]. Oxygen intercalation can be achieved by performing an annealing step in air, and it has been shown to induce a slight p-type doping [8, 79, 80]. Several activities have theoretically and experimentally addressed the effect of nitrogen intercalation [9, 81–84]. First-principles calculations suggest that nitrogen intercalation is an additional promising route to attain charge-neutral graphene [81]. Thermal treating with NH_3 has been reported to result in effective N intercalation [9], while other techniques have led to the formation of a silicon nitride passivating layer between SiC and the carbon layers [82–84]. Formation of a nitride layer at the interface has been reported to result in an increase of the graphene room-temperature mobility [85].

When the graphene lays on an ordered substrate, a coincidence lattice might form and as a consequence a superlattice potential can influence the behavior of the Dirac fermions. In Section 5.3.2 we will discuss the resulting band structure of a system obtained by intercalating a single layer of Cu atoms beneath the EG on SiC(0001) [86]. The effects of superlattice electron–phonon coupling in heavily n-doped Li-decorated graphene was predicted to give rise to a superconducting transition in graphene [87], and recently ARPES experiments gave evidence in favor of such an effect [88], although there is still controversy regarding the topic discussed in Ref. [89]. Finally, the possibility to induce or enhance a spin-orbit coupling in the graphene band structure was predicted and even measured [90–92]. This mechanism is supposed to open a gap at the Dirac point, although so far it remains experimentally elusive [93].

5.3.1 The Intercalation of Ge Atoms at the Graphene/ SiC Interface

The definition of adjacent p- and n-doped regions in semiconductors without the need of external gating can be considered one of the major accomplishments of the twentieth century solid-state physics. The p/n junction has indeed become a fundamental concept in modern electronics. The possibility to realize such a structure on a purely 2D crystal without external gating has recently gained increasing attention. Apart from the applicability as an electronic element, when produced on a graphene sheet, a lateral p/n junction was predicted to be the source of exciting physical phenomena like Klein tunneling [74, 94, 95] or electron focusing [96] (Veselago lens [97]). The concept of extrinsic doping is actually something not directly transferable from a 3D to a 2D material, exactly because of the lack of a 3D bulk matrix. However, it was observed that the adsorption of molecules on top of graphene [14, 34] and the modification of the interface via intercalation of foreign atomic species underneath graphene [1–4, 26, 68, 69, 78, 98, 99], are efficient methods to alter the carrier concentration in graphene.

In this section, the effects induced by the intercalation of nonmetallic atoms at the interface between SiC(0001) and a graphene layer will be discussed. For this work, a few monolayers of Ge were deposited on a ZLG sample and then intercalated upon annealing at 720°C in UHV. Depending on the preparation conditions, different doping could be achieved [3].

In particular, it was observed [3] that deposition of about 5 ML of Ge on graphene yielded p-doped Ge-intercalated EG. The ARPES spectrum corresponding to this p-phase is displayed in Fig. 5.11a. The carrier density extracted from the spectrum is $p \simeq 4.1 \times 10^{12}$ cm^{-2}. All three ARPES spectra of Fig. 5.11 were measured at the $\overline{K}$-point, perpendicularly to the $\overline{\Gamma K}$-direction with linearly p-polarized light of energy 90 eV. The spectrum in Fig. 5.11a proves that the buffer layer was relieved from the interaction with the substrate and the homogeneity of the intercalation process over a field of view (FOV) of 15 μm is visible in the LEEM micrograph shown in Fig. 5.11d, where an image of the p-phase was recorded at an electron beam energy of 5.3 eV. Indeed, no marked grayscale variations are observed in the image, except for the step edges. By increasing the annealing temperature, the Ge starts to sublimate from the interface [3] as visible from Fig. 5.11g, where the core-level Ge 3d peak is shown as a function of the annealing temperature. Note that such a process does not initiate at the step edge, but rather in the middle of the crystal terrace, as visible from the LEEM image in Fig. 5.11e (same region, FOV and energy as in panel [d]). When the annealing temperature reaches 920°C, about one half of the intercalated Ge is sublimated (cf Fig. 5.11g), and the graphene exhibits a fully developed n-phase, as the ARPES spectrum in Fig. 5.11c confirms. The n-phase has an estimated carrier density of $n \simeq 4.8 \times 10^{12}$ cm^{-2} [3]. The most interesting situation occurs in the intermediate stage, when the annealing temperature is raised up to 820°C. At this stage, only about one-fourth of the total Ge was sublimated (see Fig. 5.11g) and the system presents n- and p-phase regions in about equal amounts. The ARPES spectrum of Fig. 5.11b

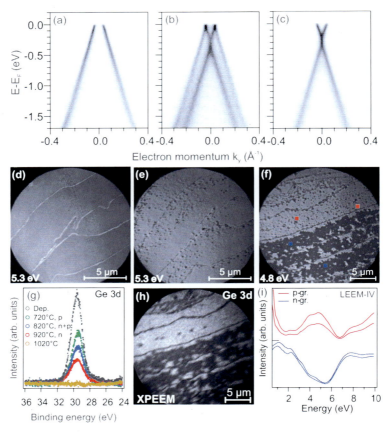

Figure 5.11 ARPES spectrum acquired along the k_y direction of a p-phase (a), mixed phase (b), and n-phase graphene after Ge intercalation. (d) LEEM micrograph of a Ge-intercalated graphene in the pure p-phase, corresponding to (a). (e) LEEM image of the initial formation of the n-phase. (f) LEEM image of the mixed p-n phase, corresponding to the ARPES spectrum of (b). (g) Ge 3d core-level peak intensity as a function of annealing temperature. The p-phase shows about double the amount of Ge as compared to the n-phase. The peaks are normalized to the Si 2p peak intensities. (h) XPEEM Ge 3d intensity map of the mixed phase, as in (f). (i) LEEM *IV* curves measured on the red/blue squares on (f). Adapted (figure) with permission from Ref. [3]. Copyright (2011) by the American Physical Society.

describes such a situation by showing the band dispersions of the Ge-intercalated graphene recorded upon annealing the sample at 820°C. In Fig. 5.11f a LEEM image on a 20 μm FOV taken at 4.8 eV visualizes the very high density of interfaces between the p and n regions at this stage. The difference in the amount of intercalated Ge between the two phases is confirmed also by X-ray photoelectron emission microscopy (XPEEM) measurements. Figure 5.11h shows an XPEEM intensity map obtained by recording the photoelectrons emitted from the Ge 3d core level. The regions visible on the image can be directly compared with what is seen in Fig. 5.11f: a higher signal (lighter gray) corresponds to a higher amount of Ge. Concerning the location of the Ge atoms in such a system, theoretical studies were carried out [100]. A stable n-doped phase is found for a Ge layer intercalated at the interface, which also provides the decoupling of the graphene from the SiC substrate. Within the model utilized in Ref. [100] however, no conditions for a stable p-phase are found and the configuration with Ge adsorbed on the graphene surface is found to be unlikely, suggesting, as speculated in [3], that the p-phase is formed when two layers of Ge are intercalated underneath graphene. Such a situation would be confirmed also by the LEEM *IV* reflectivity curves (Fig. 5.11i) which exhibit two minima for the p-phase but one for the n-phase.

5.3.2 Electronic Spectrum of a Graphene Superlattice Induced by Intercalation of Cu Atoms

Intercalation of a single-atom-thick copper layer beneath monolayer graphite was reported already several years ago on Ni(111) [101] and Ir(111) [102]. No band structure analysis was carried out though before 2016 [86]. A Cu monolayer can be intercalated at the heterointerface between the carbon-rich $(6\sqrt{3} \times 6\sqrt{3})R30°$ reconstruction and SiC(0001), by annealing the sample in UHV at temperatures close to 700°C, after a few Cu monolayers (~5 ML) were deposited from a Knudsen cell onto the surface.

The interfacial Cu monolayer is ordered in a Cu(111) fashion, aligned with the graphene. The lattice mismatch between the Cu layer and graphene gives rise to a geometrical coincidence superlattice, often referred to as a moiré pattern, which can be identified from the low-energy electron diffraction (LEED) pattern of the sample surface, as shown in Fig. 5.12a. The reciprocal lattice vectors of the SiC(0001) surface and of the graphene are indicated by arrows labelled with **s** and **g**, respectively. Hexagons of sharp satellite spots surrounding the graphene diffraction spots correspond to the long-range superstructure imposed on the graphene lattice by the new heterointerface graphene/Cu/SiC(0001). The period of this lateral superlattice as deduced from its reciprocal vectors (see the inset in Fig. 5.12a) corresponds to (13 × 13) graphene unit cells or a supercell side length of 3.2 nm. The relative dimensions and orientation of the superlattice with respect to graphene and SiC(0001) are listed in Table 5.1.

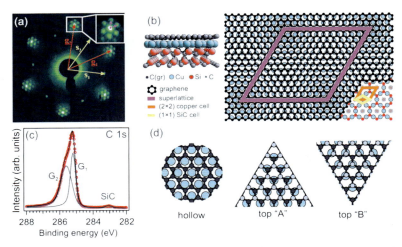

Figure 5.12 (a) LEED pattern of the epitaxial (13×13) GSL formed on the Cu/SiC(0001) interface. The reciprocal lattice vectors of the graphene and SiC lattices are indicated as (g_1, g_2) and (s_1, s_2), respectively. The inset in (a) indicates the reciprocal lattice vectors of the superlattice (m_i in the text). (b) Sketch of the corresponding atomic structure of the (13×13) GSL (neglecting possible vertical displacements) in side view (left) and top view (right) (the superlattice cell is depicted by the purple diamond). The structure of the underlying Cu/SiC interface is exposed in the lower-right corner of the top view, where the Si atoms covered by Cu atoms are indicated as shining (brownish) through the semitransparent Cu atoms. (c) C 1s core-level spectrum of the GSL taken at a photon energy of 350 eV. Experimental data are shown as open (red) circles, fitted by the solid (black) line. The fitted components G_1 and G_2 are also shown (gray lines). (d) Zoom-in of regions within the GSL unit cell of (b), with different registry between Cu atoms and the graphene unit cell, marking the distinction between hollow (Cu under C-hexagon), top A (A atoms on top), and top B (B atoms on top) configurations. From Ref. [86]. @ IOP Publishing. Reproduced with permission. All rights reserved.

Table 5.1 Summary of the mutual periodicity between the substrate, copper, graphene, and the moiré unit cells

Layer	Lattice parameter (Å)	Periodicity		Model
		Moiré	Substrate	
Graphene	2.46	13	(13×13) Gr on (12×12)Cu	
Cu	2.665	12	(2×2) Cu on $(\sqrt{3} \times \sqrt{3})$ $R30°$SiC	
SiC	3.08	$6\sqrt{3}$	1	

Notably, the unit vectors of the supercell are aligned with those of the graphene lattice so that only a single superlattice domain is present. Recently, the appearance of this moiré superstructure after Cu intercalation was also observed by scanning tunneling microscopy (STM) on a local scale [103].

In contrast with the growth of graphene on single-crystal copper, which yields polycrystalline films [104–106], the intercalation does not introduce any measurable rotational disorder, meaning that graphene maintains its epitaxial registry with the underlying SiC substrate. The superlattice emerges because of the lattice mismatch between the graphene layer and the underlying Cu/SiC interface. The copper atoms order into a well-defined epitaxial layer of hexagonal symmetry at the interface, as sketched in Fig. 5.12b. The reciprocal lattice vectors of the (moiré) superlattice are defined as $|\mathbf{m}| = m_i = g_i\text{-}c_i$, where $i = 1,2$ and g_i and c_i are the reciprocal lattice vectors of graphene and the interfacial Cu layer, respectively. The length of the Cu reciprocal lattice vector can be inferred from the

LEED measurements to be $c_i = 12/13 \cdot g_i$, which corresponds to a lattice constant of the interfacial copper layer of $a_{Cu/SiC} = 2.66$ Å. This value is 4% larger than that of the Cu(111) surface. The strain arises since the Cu atoms have to accommodate for the periodic potential of the underlying SiC substrate. In this way the Cu layer is matched epitaxially to the SiC surface so that a (2 × 2) Cu cell coincides with the $(\sqrt{3} \times \sqrt{3})R30°$ cell of SiC(0001) (see the bottom-right part in Fig. 5.12b). The presence of this interfacial order with $(\sqrt{3} \times \sqrt{3})R30°$ periodicity is indeed apparent in LEED data taken at higher energies [86]. The (13 × 13) periodicity of graphene is equivalent to a $(6\sqrt{3} \times 6\sqrt{3})R30°$ cell in terms of the SiC lattice, which is the common graphene/SiC(0001) reconstruction [11, 107], implying that no strain of the graphene layer is required. The graphene/Cu/SiC coincidence superlattice contains different local registries within the large unit cell. In some regions the copper atoms are located beneath the center of the graphene hexagons (hollow sites) or elsewhere right below one of the carbon atoms (A or B atoms, top sites), as indicated by the three sections of the supercell shown in Fig. 5.12d. A periodic modulation potential superimposed on the honeycomb lattice on a length scale shorter than the mean free path of the Dirac fermions is what is called a graphene superlattice (GSL) [108, 109]. In this case, the superlattice potential originates from the different binding energies of carbon atoms within the different local registries. This difference is reflected in the visibly broadened C 1s core-level peak of the GSL in Fig. 5.12c, which can be well fitted with two components G_1 and G_2 with a mutual chemical shift of 350 meV. Note that there is a large net shift toward higher binding energies of the graphene-related peaks as compared to neutral graphene at 284.4 eV, which reflects the strong n-doping of the graphene (see Fig. 5.13). The areal ratio of the G_1 and G_2 components amounts to about 2:3 and is directly proportional to the number of carbon atoms within the two local registries in the GSL supercell (see Fig. 5.12d). The regions represented by the peak G_2 are bound stronger to the Cu/SiC interface than those represented by the G_1 component. Stronger binding is accompanied by a shorter local adsorption length, and it can be, therefore, assumed that the resulting graphene layer is corrugated [110].

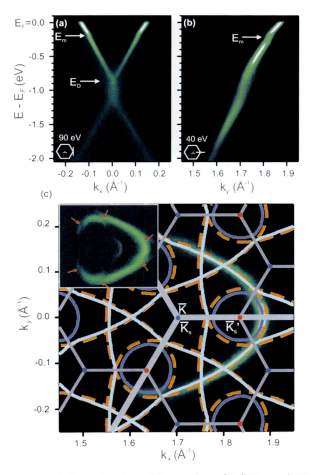

Figure 5.13 ARPES dispersion plots of the graphene/Cu/SiC superlattice taken (a) perpendicular to $\overline{\Gamma K}$ with a photon energy 90 eV and (b) along $\overline{\Gamma K}$ with a photon energy 40 eV. In both cases the light is p-polarized. (c) Experimental Fermi surface (FS) of the GSL. The thin and thick gray hexagons represent the Brillouin zone boundaries of the (13×13) superlattice and of the unreconstructed (1×1) graphene, respectively. Fits of the experimental data are plotted by light- and dark-blue (gray) lines. Dark-blue (gray) circles form at the $\overline{K_S}'$-points of the mini Brillouin zone. The dashed orange (gray) lines are the calculated Fermi contours of the unperturbed graphene displaced from the primary cone by the reciprocal lattice vectors **m** of the superlattice. The inset shows the same FS within the same range in reciprocal space, plotted with a logarithmic intensity scale. The split positions are indicated by the red arrows. From Ref. [86]. @ IOP Publishing. Reproduced with permission. All rights reserved.

The electronic band structure of the intercalated GSL sample is shown in Fig. 5.13a,b by $E(k)$-dispersion plots measured around the $\overline{K}$-point perpendicular to and along the $\overline{\Gamma K}$- azimuth, respectively. The graphene π-bands are highly electron doped due to charge transfer from the Cu/SiC interface. The Fermi level in the GSL is raised with respect to the Dirac point by about 0.85 eV, corresponding to a charge transfer of 0.05 electrons per graphene unit cell. No bandgap is observed around the Dirac point, as it can be seen in Fig. 5.13a,b, differently from what was observed for EG on Cu(111) [105, 106]. The sections with different EG/Cu stacking within the unit cell (cf Fig. 5.12d) are small enough that there is no electronic A-/B-sublattice symmetry breaking effect. However, a substantial renormalization of the energy dispersion can be observed in the vicinity of E_D with a diamond-like shape of the spectrum (cf Fig. 5.13a). This effect is very likely due to coupled photohole-plasmon modes, that is, plasmarons, as already shown for the case of K-doped H-intercalated graphene on SiC [69]. Moreover, at energies around 200 meV (labeled as E_m in Fig. 5.13b) the band velocity is further renormalized.

The Fermi surface of the GSL in the vicinity of the $\overline{K}$-valley is shown in Fig. 5.13c. In stark contrast to the continuous Fermi vector contour expected from the band structure of graphene, the Fermi surface is split along certain directions, more clearly visible in the panel's inset plotted with a logarithmic intensity grayscale and indicated by the red arrows. These split positions coincide with the mini Brillouin zone (mBZ) boundaries of the superlattice potential (thin gray hexagons). The quality and sharpness of the measured data allow one to determine the exact position of the bands in the experimental Fermi surface by fitting the raw data with a lorentzian profile at every point along the radial coordinate around the $\overline{K}$-point (cf Ref. [26, 111] for more details). The photoemission intensities of graphene's constant energy surfaces (CES) show a very strong angular dependence [56, 112], so that not all states can be evaluated with this fitting procedure. Yet, in the (13 × 13) periodicity the $\overline{K}$-point of the primary BZ (indicated by the thick gray lines) is mapped onto the $\overline{K}_S$-points of the mBZ (blue points).

As a consequence, the π-band's Dirac cone will be repeated by band folding in the corresponding repeated zone scheme [109]. Accordingly, the experimental Fermi surface can be completed by folding and threefold symmetrization with respect to the mBZ boundaries. The resulting fitted and folded points are connected by the thin cyan solid lines in Fig. 5.13c. The minicone's Fermi surface emerges as three blue circles centered at the $\overline{K_S}'$-points (red points), which are separated from the remaining Fermi lines of the original Dirac cone by about 0.01 Å^{-1} at the crossing with the mBZ boundaries. The radius of the minicones (at E_F) is about 0.05 Å^{-1}. Altogether, the contours of the minicones at E_F and the Fermi lines belonging to the main Dirac cone form the Fermi surface of the GSL. The small circle visible centered in $\overline{K}$ in the inset of Fig. 5.13c is due to a small contribution of BLG and does not arise from the presence of the lateral superlattice potential [86]. The Fermi surfaces of the replica cones as calculated by using the tight binding (TB) 3rd-NNN model described in Ref. [113] and repeated in the (13 × 13) periodicity are superimposed (dashed red lines) onto the raw data and the fitted band positions. The crossing points of the calculated states at the mBZ boundaries are indeed the positions where the experimental Fermi surfaces display their split. The Bragg reflection of the electronic states at the mBZ boundaries induces a prohibited energy region (or gap) that is proportional to the strength of the periodic potential. In Ref. [86] the superlattice potential strength was extracted from the experimental data and its mean value was estimated to be ~70 meV.

5.4   Conclusive Remarks

Epitaxial graphene (EG) on SiC remains one of the most promising platforms for the upcoming graphene-based electronics. In this chapter we have shown that via intercalation of foreign atomic species at the heterointerface between the carbon-rich (6$\sqrt{3}$ × 6$\sqrt{3}$) $R30°$ reconstruction of the SiC(0001) and the substrate itself, it is possible to selectively modify the electronic properties of EG and induce a wide range of effects that alter the properties of EG. The carrier type and carrier concentration can be engineered by choice and amount of the intercalated species. Chemically gated mesoscopic p/n junctions are accessible. Furthermore, in some circumstances the intercalants are observed to form a 2D ordered interfacial layer, thereby imposing a periodic superlattice potential that profoundly modifies the energy dispersion of the Dirac fermions in graphene.

On the other hand, graphene on top of another 2D material is by definition what is called a 2D heterostack, which recently attracted lots of attention by virtue of its potential applications [114, 115].

The quest to find a valid candidate for the postsilicon electronics era has been faced with a major disappointment because of the realization that in spite of outperforming Si in every measurable figure, graphene is a semimetal and, therefore, not usable, as it is, as a Si replacer for logic electronics [116]. We have shown though that BLG and TLG do exhibit a relevant and tunable bandgap at the Dirac point. It has been observed by means of ARPES measurements that epitaxial BLG on SiC(0001) suffers from the coexistence of massless and massive fermions [117], that is of AA and AB stackings, and therefore its density of states never shows a gap, at least on an area of about 1 mm^2. No comparable studies on high-quality TLG have been carried out so far, but we have shown that the highest contribution for the TLG is given by the ABC stacking, which makes TLG one of the most promising paths to follow.

As a future perspective we note that the mechanical patterning of graphene into 1D graphene nanoribbons has shown also very promising results in terms of bandgap and scalability [118].

References

1. Riedl, C., Coletti, C., Iwasaki, T., Zakharov, A. A., and Starke, U., Quasi-freestanding epitaxial graphene on SiC obtained by hydrogen intercalation, *Phys. Rev. Lett.*, **103**, 246804 (2009).

2. Gierz, I., Suzuki, T., Weitz, R. T., Lee, D. S., Krauss, B., Riedl, C., Starke, U., Höchst, H., Smet, J. H., Ast, C. R., and Kern, K., Electronic decoupling of an epitaxial graphene monolayer by gold intercalation, *Phys. Rev. B*, **81**, 235408 (2010).

3. Emtsev, K. V., Zakharov, A. A., Coletti, C., Forti, S., and Starke, U., Ambipolar doping in quasi-free epitaxial graphene on SiC(0001) controlled by Ge intercalation, *Phys. Rev. B*, **84**, 125423 (2011).

4. Virojanadara, C., Watcharinyanon, S., Zakharov, A. A., and Johansson, L. I., Epitaxial graphene on 6*H* -SiC and Li intercalation, *Phys. Rev. B*, **82**, 205402 (2010).

5. Watcharinyanon, S., Johansson, L., Xia, C., and Virojanadara, C., Changes in structural and electronic properties of graphene grown on

6H-SiC(0001) induced by Na deposition, *J. Appl. Phys.*, **111**, 083711 (2012).

6. Sandin, A., Jayasekera, T., Rowe, J., Kim, K., Buongiorno Nardelli, M., and Dougherty, D., Multiple coexisting intercalation structures of sodium in epitaxial graphene-SiC interfaces, *Phys. Rev. B*, **85**, 125410 (2012).

7. Xia, C., Watcharinyanon, S., Zakharov, A., Yakimova, R., Hultman, L., Johansson, L., and Virojanadara, C., Si intercalation/deintercalation of graphene on 6H-SiC(0001), *Phys. Rev. B*, **85**, 045418 (2012).

8. Oliveira, M. H. J., Schumann, T., Fromm, F., Koch, R., Ostler, M., Ramsteiner, M., Seyller, T., Lopes, J. M. J., and Riechert, H., Formation of high-quality quasi-free-standing bilayer graphene on SiC(0001) by oxygen intercalation upon annealing in air, *Carbon*, **52**, 83 (2013).

9. Wang, Z.-J., Wei, M., Jin, L., Ning, Y., Yu, L., Fu, Q., and Bao, X., Simultaneous N-intercalation and N-doping of epitaxial graphene on 6H-SiC(0001) through thermal reactions with ammonia, *Nano Res.*, **6**, 399 (2013).

10. Riedl, C., Starke, U., Bernhardt, J., Franke, M., and Heinz, K., Structural properties of the grapheme SiC(0001) interface as a key for the preparation of homogeneous large-terrace graphene surfaces, *Phys. Rev. B*, **76**, 245406 (2007).

11. Emtsev, K. V., Speck, F., Seyller, T., Ley, L., and Riley, J. D., Interaction, growth, and ordering of epitaxial graphene on SiC{0001} surfaces: a comparative photoelectron spectroscopy study, *Phys. Rev. B*, **77**, 155303 (2008).

12. Goler, S., Coletti, C., Piazza, V., Pingue, P., Colangelo, F., Pellegrini, V., Emtsev, K. V., Forti, S., Starke, U., Beltram, F., and Heun, S., Revealing the atomic structure of the buffer layer between SiC(0001) and epitaxial graphene, *Carbon*, **51**, 249 (2013).

13. Mallet, P., Varchon, F., Naud, C., Magaud, L., Berger, C., and Veuillen, J.-Y., Electron states of mono- and bilayer graphene on SiC probed by scanning-tunneling microscopy, *Phys. Rev. B*, **76**, 041403 (2007).

14. Starke, U., Forti, S., Emtsev, K., and Coletti, C., Engineering the electronic structure of epitaxial graphene by transfer doping and atomic intercalation, *MRS Bull.*, **37**, 1177 (2012).

15. Soubatch, S., Saddow, S. E., Rao, S. P., Lee, W., Konuma, M., and Starke, U., Structure and morphology of 4H-SiC wafer surfaces after H_2-etching, *Mater. Sci. Forum*, **483–485**, 761 (2005).

16. Frewin, C. L., Coletti, C., Riedl, C., Starke, U., and Saddow, S. E., A Comprehensive study of hydrogen etching on the major SiC polytypes and crystal orientations, *Mater. Sci. Forum*, **615–617**, 589 (2009).

17. Riedl, C., Zakharov, A. A., and Starke, U., Precise *in situ* thickness analysis of epitaxial graphene layers on SiC(0001) using low-energy electron diffraction and angle resolved ultraviolet photoelectron spectroscopy, *Appl. Phys. Lett.*, **93**, 033106 (2008).

18. Emtsev, K. V., Bostwick, A., Horn, K., Jobst, J., Kellog, G. L., Ley, L., McChesney, J., Ohta, T., Reshanov, S. A., Röhrl, J., Rotenberg, E., Schmid, A., Waldmann, D., Weber, H. B., and Seyller, T., Towards wafer-size graphene layers by atmospheric pressure graphitization of silicon carbide, *Nat. Mater.*, **8**, 203 (2009).

19. Coletti, C., Forti, S., Principi, A., Emtsev, K. V., Zakharov, A. A., Daniels, K. M., Daas, B. K., Chandrashekhar, M. V. S., Ouisse, T., Chaussende, D., MacDonald, A. H., Polini, M., and Starke, U., Revealing the electronic band structure of trilayer graphene on SiC: an angle-resolved photoemission study, *Phys. Rev. B*, **88**, 155439 (2013).

20. Tsuchida, H., Kamata, I., and Izumi, K., Infrared attenuated total reflection spectroscopy of 6H-SiC(0001) and (0001) surfaces, *J. Appl. Phys.*, **85**, 3569 (1999).

21. Seyller, T., Passivation of hexagonal SiC surfaces by hydrogen termination, *J. Phys.: Condens. Matter*, **16**, S1755 (2004).

22. Coletti, C., Frewin, C., Hoff, H. M., and Saddow, S. E., Electronic passivation of 3C-SiC(001) via hydrogen treatment semiconductor devices, materials, and processing, *Electrochem. Solid-State Lett.*, **11**, H285 (2008).

23. Ohta, T., Bostwick, A., Seyller, T., Horn, K., and Rotenberg, E., Controlling the electronic structure of bilayer graphene, *Science*, **313**, 951 (2006).

24. Zhou, S. Y., Gweon, G.-H., Fedorov, A. V., First, P. N., de Heer, W. A., Lee, D.-H., Guinea, F., Castro Neto, A. H., and Lanzara, A., Substrate-induced bandgap opening in epitaxial graphene, *Nat. Mater.*, **6**, 770 (2007).

25. Geim, A. K., Graphene: status and prospects, *Science*, **324**, 1530 (2009).

26. Forti, S., Emtsev, K. V., Coletti, C., Zakharov, A. A., Riedl, C., and Starke, U., Large-area homogeneous quasi-free standing epitaxial graphene on SiC(0001): electronic and structural characterization, *Phys. Rev. B*, **84**, 125449 (2011).

27. Walter, A. L., Bostwick, A., Jeon, K.-J., Speck, F., Ostler, M., Seyller, T., Moreschini, L., Chang, Y. J., Polini, M., Asgari, R., MacDonald, A. H., Horn, K., and Rotenberg, E., Effective screening and the plasmaron bands in graphene, *Phys. Rev. B*, **84**, 085410 (2011).

28. Ristein, J., Mammadov, S., and Seyller, T., Origin of doping in quasi-free-standing graphene on silicon carbide, *Phys. Rev. Lett.*, **108**, 246104 (2012).

29. Varchon, F., Mallet, P., Veuillen, J.-Y., and Magaud, L., Ripples in epitaxial graphene on the Si-terminated SiC(0001) surface, *Phys. Rev. B*, **77**, 235412 (2008).

30. Hibino, H., Kageshima, H., Maeda, F., Nagase, M., Kobayashi, Y., and Yamaguchi, H., Microscopic thickness determination of thin graphite films formed on SiC from quantized oscillaton in reflectivity of low-energy electrons, *Phys. Rev. B*, **77**, 075413 (2008).

31. Sieber, N., Seyller, T., Ley, L., James, D., Riley, J. D., Leckey, R. C. G., and Polcik, M., Synchrotron X-ray photoelectron spectroscopy study of hydrogen-terminated 6H-SiC{0001} surfaces, *Phys. Rev. B*, **67**, 205304 (2003).

32. McCann, E., Asymmetry gap in the electronic bandstructure of bilayer graphene, *Phys. Rev. B*, **74**, 161403 (2006).

33. Castro, E. V., Novoselov, K. S., Morozov, S. V., Peres, N. M. R., dos Santos, J. M. B. L., Nilsson, J., Guinea, F., Geim, A. K., and Castro Neto, A. H., Biased bilayer graphene: semiconductor with a gap tunable by the electric field effect, *Phys. Rev. Lett.*, **99**, 216802 (2007).

34. Coletti, C., Riedl, C., Lee, D. S., Krauss, B., Patthey, L., von Klitzing, K., Smet, J. H., and Starke, U., Charge neutrality and band-gap tuning of epitaxial graphene on SiC by molecular doping, *Phys. Rev. B*, **81**, 235401 (2010).

35. Wang, T. H., Zhu, Y. F., and Jiang, Q., Bandgap opening of bilayer graphene by dual doping from organic molecule and substrate, *J. Phys. Chem. C*, **117**, 12873 (2013).

36. Zhang, Y., Tang, T.-T., Girit, C., Hao, Z., Martin, M. C., Zettl, A., Crommie, M. F., Shen, Y. R., and Wang, F., Direct observation of a widely tunable bandgap in bilayer graphene, *Nature*, **459**, 820 (2009).

37. Samuels, A. J., and Carey, J. D., Molecular doping and band-gap opening of bilayer graphene, *ACS Nano*, **7**, 2790 (2013).

38. Chen, W., Chen, S., Qi, D. C., Gao, X. Y., and Wee, A. T. S., Surface transfer p-type doping of epitaxial graphene, *J. Am. Chem. Soc.* **129**, 10418 (2007).

39. Min, H., and MacDonald, A. H., Electronic structure of multilayer graphene, *Prog. Theor. Phys. Suppl.*, **176**, 227 (2008).

40. Zhang, F., Sahu, B., Min, H., and MacDonald, A. H., Band structure of ABC-stacked graphene trilayers, *Phys. Rev. B*, **82**, 035409 (2010).

41. Koshino, M., Interlayer screening effect in graphene multilayers with *ABA* and *ABC* stacking, *Phys. Rev. B*, **81**, 125304 (2010).

42. Craciun, M. F., Russo, .. Yamamoto, M., Oostinga, J. B., Morpurgo, A. F., and Tarucha, S., Trilayer graphene is a semimetal with a gate-tunable band overlap, Nat. Inotech., **4**, 383 (2009).

43. Lui, C. H., Li, Z., Mak, K. F., Cappelluti, E., and Heinz, T. F., Observation of an electrically tunable band gap in trilayer graphene, Nat. Phys., **7**, 944 (2011).

44. Bao, W., Jing, L., Er, J. V., Lee, Y., Liu, G., Tran, D., Standley, B., Aykol, M., Cronin, S. B., Smirnov, D., Koshino, M., McCann, E., Bockrath, M., and Lau, C. N., Stacking-dependent band gap and quantum transport in trilayer graphene, Nat. Phys., **7**, 948 (2011).

45. Zhang, L., Zhang, Y., Camacho, J., Khodas, M., and Zalinzyak, I., The experimental observation of quantum Hall effect I=3 chiral quasiparticles in trilayer graphene, Nat. Phys., **7**, 953 (2011).

46. Yacoby, A., Tri and tri again, Nat. Phys., **7**, 925 (2011).

47. Guinea, F., Castro Neto, A. H., and Peres, N. M. R., Electronic states and Landau levels in graphene stacks, Phys. Rev. B, **73**, 245426 (2006).

48. Aoki, M., and Amawashi, H., Dependence of band structures on stacking and field in layered graphene, Solid State Commun., **142**, 123 (2007).

49. Grüneis, A., Attaccalite, C., Wirtz, L., Shiozawa, H., Saito, R., Pichler, T., and Rubio, A., Tight-binding description of the quasiparticle dispersion of graphite and few-layer graphene, Phys. Rev. B, **78**, 205425 (2008).

50. Koshino, M., and McCann, E., Trigonal warping and Berry's phase $N\pi$ in ABC-stacked multilayer graphene, Phys. Rev. B, **80**, 165409 (2009).

51. Avetisyan, A. A., Partoens, B., and Peeters, F. M., Stacking order dependent electric field tuning of the band gap in graphene multilayers, Phys. Rev. B, **81**, 115432 (2010).

52. Lipson, H., and Stokes, A. R., The structure of graphite, Proc. R. Soc. A, **101**, 181 (1942).

53. Lui, C. H., Li, Z., Chen, Z., Klimov, P. V., Brus, L. E., and Heinz, T. F., Imaging stacking order in few-layer graphene, Nano Lett., **11**, 164 (2011).

54. Ohta, T., Bostwick, A., McChesney, J. L., Seyller, T., Horn, K., and Rotenberg, E., Interlayer interaction and electronic screening in multilayer graphene investigated with angle-resolved photoemission spectroscopy, Phys. Rev. Lett., **98**, 206802 (2007).

55. McCann, E., and Fal'ko, V. I., Landau-level degeneracy and quantum Hall effect in a graphite bilayer, Phys. Rev. Lett., **96**, 086805 (2006).

56. Mucha-Kruczyński, M., Tsyplyatyev, O., Grishin, A., McCann, E., Fal'ko, V. I., Bostwick, A., and Rotenberg, E., Characterization of graphene

through anisotropy of constant-energy maps in angle-resolved photoemission, *Phys. Rev. B*, **77**, 195403 (2008).

57. Norimatsu, W., and Kusunoki, M., Selective formation of ABC-stacked graphene layers on SiC(0001), *Phys. Rev. B*, **81**, 161410 (2010).

58. Saddow, S., Coletti, C., Frewin, C., Schettini, N., Oliveros, A., and Jarosezski, M., Single-crystal silicon carbide: a biocompatible and hemocompatible semiconductor for advanced biomedical applications, *Mater. Res. Soc. Symp. Proc.*, **1246**, 193 (2010).

59. Ouerghi, A., Marangolo, M., Belkhou, R., El Moussaoui, S., Silly, M. G., Eddrief, M., Largeau, L., Portail, M., Fain, B., and Sirotti, F., Epitaxial graphene on 3C-SiC(111) pseudosubstrate: structural and electronic properties, *Phys. Rev. B*, **82**, 125445 (2010).

60. Ouerghi, A., Belkhou, R., Marangolo, M., Silly, M. G., El Moussaoui, S., Eddrief, M., Largeau, L., Portail, M., and Sirotti, F., Structural coherency of epitaxial graphene on 3C-SiC(111) epilayers on Si(111), *Appl. Phys. Lett.*, **97**, 161905 (2010).

61. Chaussende, D., Latu-Romain, L., Auvray, L., Ucar, M., Pons, M., and Madar, R., Large area DPB free (111) β-SiC thick layer grown on (0001) α-SiC nominal surfaces by the CF-PVT method, *Mater. Sci. Forum*, **483–485**, 225 (2005).

62. Coletti, C., Emtsev, K. V., Zakharov, A. A., Ouisse, T., Chaussende, D., and Starke, U., Large area quasi-free standing monolayer graphene on 3C-SiC(111), *Appl. Phys. Lett.*, **99**, 081904 (2011).

63. Bolotin, K., Sikes, K., Jiang, Z., Klima, M., Fudenberg, G., Hone, J., Kim, P., and Stormer, H., Ultrahigh electron mobility in suspended graphene, *Solid State Commun.*, **146**, 351 (2008).

64. Pallecchi, E., Lafont, F., Cavaliere, V., Schopfer, F., Mailly, D., Poirier, W., and Ouerghi, A., High electron mobility in epitaxial graphene on 4H-SiC(0001) via post-growth annealing under hydrogen, *Sci. Rep.*, **4**, 4558 (2014).

65. Tanabe, S., Takamura, M., Harada, Y., Kageshima, H., and Hibino, H., Effects of hydrogen intercalation on transport properties of quasi-free-standing monolayer graphene, *Jpn. J. Appl. Phys.*, **53**, 04EN01 (2014).

66. Waldmann, D., Jobst, J., Speck, F., Seyller, T., Krieger, M., and Weber, H. B., Bottom-gated epitaxial graphene, *Nat. Mater.*, **10**, 357 (2011).

67. Hicks, J., and Conrad, E., Graphene investigated by synchrotron radiation, *MRS Bull.*, **37**, 1203 (2012).

68. Bostwick, A., Ohta, T., Seyller, T., Horn, K., and Rotenberg, E., Quasiparticle dynamics in graphene, *Nat. Phys.*, **3**, 36 (2007).

69. Bostwick, A., Speck, F., Seyller, T., Horn, K., Polini, M., Asgari, R., MacDonald, A. H., and Rotenberg, E., Observation of plasmarons in quasi-freestanding doped graphene, *Science*, **328**, 999 (2010).

70. Robinson, J. A., Hollander, M., LaBella, M., Trumbull, K. A., Cavalero, R., and Snyder, D. W., Epitaxial graphene transistors: enhancing performance via hydrogen intercalation, *Nano Lett.*, **11**, 3875 (2011).

71. Chuang, F.-C., Lin, W.-H., Huang, Z.-Q., Hsu, C.-H., Kuo, C.-C., Ozolins, V., and Yeh, V., Electronic structures of an epitaxial graphene monolayer on SiC(0001) after gold intercalation: a first-principles study, *Nanotechnology*, **22**, 275704 (2011).

72. Hsu, C.-H., Lin, W.-H., Ozolins, V., and Chuang, F.-C., Electronic structures of an epitaxial graphene monolayer on SiC(0001) after metal intercalation (metal = Al, Ag, Au, Pt, and Pd): a first-principles study, *Appl. Phys. Lett.*, **100**, 063115 (2012).

73. Baringhaus, J., Stöhr, A., Forti, S., Krasnikov, S. A., Zakharov, A. A., Starke, U., and Tegenkamp, C., Bipolar gating of epitaxial graphene by intercalation of Ge, *Appl. Phys. Lett.*, **104**, 261602 (2014).

74. Beenakker, C. W. J., Colloquium: Andreev reflection and Klein tunneling in graphene, *Rev. Mod. Phys.* **80**, 1337 (2008).

75. Baringhaus, J., Stöhr, A., Forti, S., Starke, U., and Tegenkamp, C., Ballistic bipolar junctions in chemically gated graphene ribbons, *Sci. Rep.*, **5**, 9955 (2015).

76. Watcharinyanon, S., Johansson, L., Zakharov, A., and Virojanadara, C., Studies of Li intercalation of hydrogenated graphene on SiC(0001), *Surf. Sci.*, **606**, 401 (2012).

77. Deretzis, I., and La Magna, A., Role of covalent and metallic intercalation on the electronic properties of epitaxial graphene on SiC(0001), *Phys. Rev. B*, **84**, 235426 (2011).

78. Walter, A. L., Jeon, K.-J., Bostwick, A., Speck, F., Ostler, M., Seyller, T., Moreschini, L., Kim, Y. S., Chang, Y. J., Horn, K., and Rotenberg, E., Highly p-doped epitaxial graphene obtained by fluorine intercalation, *App. Phys. Lett.*, **98**, 184102 (2011).

79. Ostler, M., Fromm, F., Koch, R. J., Wehrfritz, P., Speck, F., Vita, H., Böttcher, S., Horn, K., and Seyller, T., Buffer layer free graphene on SiC(0001) via interface oxidation in water vapor, *Carbon*, **70**, 258 (2014).

80. Bom, N., Oliveira Jr., M., Soares, G., Radtke, C., Lopes, J., and Riechert, H., Synergistic effect of H_2O and O_2 on the decoupling of epitaxial

monolayer graphene from SiC(0001) via thermal treatments, *Carbon*, **78**, 298 (2014).

81. Caffrey, N. M., Armiento, R., Yakimova, R., and Abrikosov, I. A., Charge neutrality in epitaxial graphene on 6*H*-SiC(0001) via nitrogen intercalation, *Phys. Rev. B*, **92**, 081409 (2015).

82. Chai, J. W., Pan, J. S., Zhang, Z., Wang, S. J., Chen, Q., and Huan, C. H. A., X-ray photoelectron spectroscopy studies of nitridation on 4H-SiC(0001) surface by direct nitrogen atomic source, *Appl. Phys. Lett.*, **92**, 092119 (2008).

83. Tsai, H.-S., Lai, C.-C., Medina, H., Lin, S.-M., Shih, Y.-C., Chen, Y.-Z., Liang, J.-H., and Chueh, Y.-L., Scalable graphene synthesised by plasma-assisted selective reaction on silicon carbide for device applications, *Nanoscale*, **6**, 13861 (2014).

84. Vélez-Fort, E., Pallecchi, E., Silly, M. G., Bahri, M., Patriarche, G., Shukla, A., Sirotti, F., and Ouerghi, A., Single step fabrication of N-doped graphene/Si_3N_4/SiC heterostructures, *Nano Res.*, **7**, 835 (2014).

85. Masuda, Y., Norimatsu, W., and Kusunoki, M., Formation of a nitride interface in epitaxial graphene on SiC (0001), *Phys. Rev. B*, **91**, 075421 (2015).

86. Forti, S., Stöhr, A., Zakharov, A. A., Coletti, C., Emtsev, K. V., and Starke, U., Mini-Dirac cones in the band structure of a copper intercalated epitaxial graphene superlattice, *2D Mater.*, **3**, 035003 (2016).

87. Profeta, G., Calandra, M., and Mauri, F., Phonon-mediated superconductivity in graphene by lithium deposition, *Nat. Phys.*, **8**, 131 (2012).

88. Ludbrook, B. M., Levy, G., Nigge, P., Zonno, M., Schneider, M., Dvorak, D. J., Veenstra, C. N., Zhdanovich, S., Wong, D., Dosanjh, P., Straßer, C., Stöhr, A., Forti, S., Ast, C. R., Starke, U., and Damascelli, A., Evidence for superconductivity in Li-decorated monolayer graphene, *Proc. Natl. Acad. Sci. U S A*, **112**, 11795–11799 (2015).

89. Khademi, A., Sajadi, E., Dosanjh, P., Folk, J., Stöhr, A., Forti, S., and Starke, U., Transport measurement of Li doped monolayer graphene, in *APS Meeting Abstracts* (2016).

90. Kane, C. L., and Mele, E. J., Quantum spin hall effect in graphene, *Phys. Rev. Lett.*, **95**, 226801 (2005).

91. Weeks, C., Hu, J., Alicea, J., Franz, M., and Wu, R., Engineering a robust quantum spin Hall state in graphene via adatom deposition, *Phys. Rev. X*, **1**, 021001 (2011).

92. Avsar, A., Tan, J. Y., Taychatanapat, T., Balakrishnan, J., Koon, G., Yeo, Y., Lahiri, J., Carvalho, A., Rodin, A. S., O'Farrell, E., Eda, G., Castro Neto, A. H., and Özyilmaz, B., Spin-orbit proximity effect in graphene, *Nat. Commun.*, **5**, 4875 (2014).

93. Stöhr, A., Forti, S., Link, S., Zakharov, A., Kern, K., Starke, U., and Benia, H., Intercalation of graphene on SiC(0001) via ion-implantation, *Phys. Rev. B*, **94**, 085431 (2016).

94. Katsnelson, M. I., Novoselov, K. S., and Geim, A. K., Chiral tunnelling and the Klein paradox in graphene, *Nat. Phys.*, **2**, 620 (2006).

95. Stander, N., Huard, B., and Goldhaber-Gordon, D., Evidence for Klein tunneling in graphene *p-n* junctions, *Phys. Rev. Lett.*, **102**, 026807 (2009).

96. Cheianov, V. V., Fal'ko, V., and Altshuler, B. L., The focusing of electron flow and a veselago lens in graphene p-n junctions, *Science*, **315**, 1252 (2007).

97. Veselago, V. G., The electrodynamics of substances with simultaneously negative values of E and μ, *Sov. Phys. Usp.*, **10**, 509 (1968).

98. Gierz, I., Riedl, C., Starke, U., Ast, C. R., and Kern, K., Atomic hole doping of graphene, *Nano Lett.*, **8**, 4603 (2008).

99. McChesney, J. L., Bostwick, A., Ohta, T., Seyller, T., Horn, K., González, J., and Rotenberg, E., Extended van hove singularity and superconducting instability in doped graphene, *Phys. Rev. Lett.*, **104**, 136803 (2010).

100. Deretzis, I., and Magna, A. L., Ab initio study of Ge intercalation in epitaxial graphene on SiC(0001), *Appl. Phys. Express*, **4**, 125101 (2011).

101. Dedkov, Y. S., Shikin, A. M., Adamchuk, V. K., Molodtsov, S. L., Laubschat, C., Bauer, A., and Kaindl, G., Intercalation of copper underneath a monolayer of graphite on Ni(111), *Phys. Rev. B*, **64**, 035405 (2001).

102. Rut'kov, E., and Gall, N., Penetration (intercalation) of copper atoms under a graphene layer on iridium (111), *Semiconductors*, **43**, 1255 (2009).

103. Yagyu, K., Tajiri, T., Kohno, A., Takahashi, K., Tochihara, H., Tomokage, H., and Suzuki, T., Fabrication of a single layer graphene by copper intercalation on a SiC(0001) surface, *Appl. Phys. Lett.*, **104**, 053115 (2014).

104. Nie, S., Wofford, J. M., Bartelt, N. C., Dubon, O. D., and McCarty, K. F., Origin of the mosaicity in graphene grown on Cu(111), *Phys. Rev. B*, **84** (2011).

105. Walter, A. L., Nie, S., Bostwick, A., Kim, K. S., Moreschini, L., Chang, Y. J., Innocenti, D., Horn, K., McCarty, K. F., and Rotenberg, E., Electronic structure of graphene on single-crystal copper substrates, *Phys. Rev. B*, **84**, 195443 (2011).

106. Siegel, D. A., Hwang, C., Fedorov, A. V., and Lanzara, A., Electron-phonon coupling and intrinsic bandgap in highly-screened graphene, *New J. Phys.*, **14**, 095006 (2012).

107. Riedl, C., Coletti, C., and Starke, U., Structural and electronic properties of epitaxial graphene on SiC(0001): a review of growth, characterization, transfer doping and hydrogen intercalation, *J. Phys. D: Appl. Phys.*, **43**, 374009 (2010).

108. Park, C.-H., Yang, L., Son, Y.-W., Cohen, M. L., and Louie, S. G., Anisotropic behaviours of massless Dirac fermions in graphene under periodic potentials, *Nat. Phys.*, **4**, 213 (2008).

109. Park, C.-H., Yang, L., Son, Y.-W., Cohen, M. L., and Louie, S. G., New generation of massless Dirac fermions in graphene under external periodic potentials, *Phys. Rev. Lett.*, **101**, 126804 (2008).

110. Preobrajenski, A. B., Ng, M. L., Vinogradov, A. S., and Mårtensson, N., Controlling graphene corrugation on lattice-mismatched substrates, *Phys. Rev. B*, **78**, 073401 (2008).

111. Forti, S., Large-Area *Epitaxial Graphene on SiC(0001): From Decoupling to Interface Engineering*, PhD thesis, Erlangen-Nürnberg University (2014).

112. Shirley, E. L., Terminello, L. J., Santoni, A., and Himpsel, F. J., Brillouin-zone-selection effects in graphite photoelectron angular distributions, *Phys. Rev. B*, **51**, 13614 (1995).

113. Reich, S., Maultzsch, J., Thomsen, C., and Ordejón, P., Tight-binding description of graphene, *Phys. Rev. B*, **66**, 035412 (2002).

114. Zhang, W., Wang, Q., Chen, Y., Wang, Z., and Wee, A. T. S., Van der Waals stacked 2D layered materials for optoelectronics, *2D Mater.*, **3**, 022001 (2016).

115. Novoselov, K. S., Jiang, D., Schedin, F., Booth, T. J., Khotkevich, V. V., Morozov, S. V., and Geim, A. K., Two-dimensional atomic crystals, *Proc. Nat. Acad. Sci. U S A*, **102**, 10451 (2005).

116. Schwierz, F., Graphene transistors, *Nat. Nanotech.*, **5**, 487 (2010).

117. Kim, K. S., Walter, A. L., Moreschini, L., Seyller, T., Horn, K., Rotenberg, E., and Bostwick, A., Coexisting massive and massless Dirac fermions in symmetry-broken bilayer graphene, *Nat. Mater.*, **12**, 887 (2013).

118. Hicks, J., Tejeda, A., Taleb-Ibrahimi, A., Nevius, M. S., Wang, F., Shepperd, K., Palmer, J., Bertran, F., Fèvre, P. L., Kunc, J., de Heer, W. A., Berger, C., and Conrad, E. H., A wide-bandgap metal-semiconductor-metal nanostructure made entirely from graphene, *Nat. Phys.*, **9**, 49 (2013).

119. Coletti, C., Forti, S., Emtsev, K. V., and Starke, U. Tailoring the electronic structure of epitaxial graphene on SiC(0001): Transfer doping and hydrogen intercalation, in Ottaviano, L., and Morandi, V. (eds) *GraphITA 2011. Carbon Nanostructures*, Springer, Berlin, Heidelberg (2012).

Chapter 6

Epitaxial Graphene on SiC: 2D Sheets, Selective Growth, and Nanoribbons

C. Berger,[a,b] D. Deniz,[a] J. Gigliotti,[a] J. Palmer,[a] J. Hankinson,[a] Y. Hu,[a] J.-P. Turmaud,[a] R. Puybaret,[c] A. Ougazzaden,[c] A. Sidorov,[a] Z. Jiang,[a] and W. A. de Heer[a,d]

[a]*School of Physics, Georgia Institute of Technology, Atlanta, GA 30332, USA*
[b]*Institut Néel, CNRS- Université Grenoble Alpes, 38042 Grenoble, France*
[c]*Georgia Institute of Technology - CNRS UMI 2958, 57070 Metz, France*
[d]*Tianjin International Center for Nanoparticles and Nanosystems, Tianjin University, China*
claire.berger@neel.cnrs.fr

Epitaxial graphene grown on SiC by the confinement-controlled sublimation method is reviewed, with an emphasis on multilayer and monolayer epitaxial graphene on the carbon face of 4H-SiC and on directed and selectively grown structures under growth-arresting or growth-enhancing masks. Recent developments in the growth of templated graphene nanostructures are also presented, as exemplified by tens of microns long, very-well-confined, and isolated 20–40 nm wide graphene ribbons. A scheme for large-scale integration of ribbon arrays with a Si wafer is also presented.

Growing Graphene on Semiconductors
Edited by Nunzio Motta, Francesca Iacopi, and Camilla Coletti
Copyright © 2017 Pan Stanford Publishing Pte. Ltd.
ISBN 978-981-4774-21-5 (Hardcover), 978-1-315-18615-3 (eBook)
www.panstanford.com

6.1 Introduction

Fifteen years of research on epitaxial graphene on SiC (hereafter called epigraphene) has largely demonstrated its potential not only as the best graphene nanoelectronics platform but also as the best platform for a large variety of basic science studies of graphene [1]. From its inception in 2001, it was realized that graphene for nanoelectronics needs an atomically well-defined substrate. This requirement is satisfied by growing the material on single-crystal substrates. Previous surface science studies of single-layer graphite grown on a hexagonal SiC substrate had demonstrated that it has characteristics of isolated graphene. In general, large-scale integration calls for commercially available wafer-scale substrates and the best graphene science is demonstrated on flat, unstrained, and unrippled surfaces; all this makes SiC a substrate of choice for graphene. SiC is widely available at affordable prices, as is clear from its widespread use as a substrate for light-emitting diode (LED) lighting.

Epigraphene production methods have been developed, allowing not only 2D sheets [2–11] but also intricately patterned structures to be grown on the substrate [12, 13]. The ability to grow graphene nanostructures cannot be underestimated: if graphene is to be used in nanoelectronics, atomically defined graphene structures need to be patterned at precise locations. The methods presented here [12, 14–16] (masking and more especially template growth on the sidewall of SiC trenches) circumvent the detrimental effect of traditional plasma etching methods, and resulting nanoribbons show exceptional room-temperature ballistic transport properties [17].

Furthermore, to achieve the best performance graphene needs to be annealed and encapsulated in order to mitigate the effects of adsorbates from exposure to the ambient atmosphere. An advantage of epigraphene is that it is resistant to high temperatures and harsh radiation conditions. This review will focus on two main aspects of epigraphene material. First we will outline the quality of epigraphene 2D sheets when grown by the confinement-controlled sublimation (CCS) method; then we will present alternative methods to pattern graphene at the nanoscale, with an emphasis on structured growth, and show paths toward large-scale integration.

6.2 Near-Equilibrium Confinement-Controlled Sublimation Growth

The growth of graphene by SiC decomposition is nonconventional in the sense that it doesn't require an external source of carbon. Rather, at a high temperature and in vacuum, the SiC surfaces decompose by silicon sublimation, resulting in a carbon-rich surface that forms a graphene layer on the SiC [18]. The key to the growth of high-crystalline-quality graphene layer is to control the rate of silicon escape Γ_{Si} from the surface, which is a balance between the rate of sublimation from and absorption onto the surface (Fig. 6.1a) [3]. Γ_{Si} is controlled by both the SiC temperature and the background pressure of the silicon vapor. When the SiC surface is in equilibrium with the Si vapor ($\Gamma_{Si} = 0$), the formation of graphene is arrested.

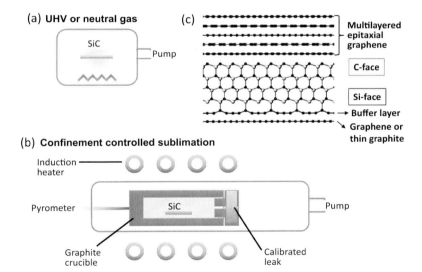

Figure 6.1 Growth of epigraphene by the confinement-controlled sublimation method. (a) Principle of graphene growth by thermal decomposition of SiC at a high temperature; Si escapes and is partially reabsorbed on the SiC surface. (b) CCS growth furnace with induction heating elements and a graphite crucible with a calibrated hole to contain the Si vapor. (c) Schematics of graphene layer growth on the top 6H-/4H-SiC(000-1) and bottom (000-1) surfaces.

Epitaxial Graphene on SiC

The CCS growth method relies on a near-equilibrium growth condition by controlling Γ_{Si}. Details can be found in Ref. [3]. A SiC crystal is placed in a graphite enclosure (the crucible). During annealing the evolving Si vapor escapes through a small calibrated hole (see Fig. 6.1b). A background of neutral gas pressure may be supplied that further slows down silicon diffusion. The graphite crucible is placed in a vacuum chamber. The system is pumped down to a base pressure of 10^{-7} mbar with a turbo-molecular pump and then uniformly heated using radio-frequency (RF) coils that heat the crucible, as shown Fig. 6.1c, to temperatures that can reach up to 2100°C. The system is designed to have no heating elements inside the vacuum chamber, so the SiC surface is exposed only to its own vapor. In our most recent furnace design [19], the induction susceptor is the graphite crucible itself. It is supported directly by the quartz tube, whose temperature remains well below its melting point even for the highest crucible temperature (2100°C for more than 20 min), owing to the small actual contact area between the tube and the crucible. The compact design allows for fast pumping speed. The temperature is measured externally with a two-wavelength pyrometer, and the assembly has very little thermal inertia, so fast heating and cooling rates up to 150°C/s can be achieved in the whole temperature range. The temperature profile is completely computer automated, with temperature ramps and plateaus fully configurable by the user. The RF power is adjusted using a proportional-integral-derivative controller that manages the feedback loop between the measured and set temperatures. We have estimated [19] that gas pressure equilibration is reached in the crucible within 5 μs, which is many orders of magnitude shorter than a typical growth time. This confirms that growth occurs very close to equilibrium. Further, since solid/vapor pressure is an intensive property that does not scale with surface area when in equilibrium, the SiC chip area inside the crucible does not affect the growth rate.

To grow graphene on hexagonal 4H- or 6H-SiC wafers a typical temperature cycle includes a degasing stage at 750°C–800°C for a few minutes, followed by a SiC surface structuring stage around 1100°C–1250°C before Si sublimation at high temperature (1400°C–1650°C). The optimum temperature and growth times are empirically determined, and stable recipes have been established for each type of structure, including the graphene buffer layer, monolayer, and bilayer on the Si face, from a monolayer thick to more than 50 layers thick on the C face (see Fig. 6.1d), and for confined growth on other than on-axis facets (e.g., SiC sidewalls; see later). The exact temperature profile depends on the background pressure (vacuum or 1 atm Ar) and the geometry (explicitly the hole size) and condition of the graphite crucible. We have noted a slow evolution with time of the growth parameters, which is due to a change in the adsorption of Si on graphite crucible walls after many growth cycles. The original conditions can be restored by baking the graphite crucible at a high temperature.

6.2.1 Multilayer C Face

The near-equilibrium growth condition of the CCS method is favorable to grow very high-quality epigraphene films. On the 4H-/6H-SiC carbon face, typically 5- to 10-layer films are easily grown [20] and more than 40 layers can be produced [21], while single-layer films are much harder to achieve [22]. The number of layers is determined by both temperature (in the range 1450°C–1525°C) and time (1 min to several 1 h cycles for very thick films).

Figure 6.2 presents atomic force microscopy (AFM) and scanning tunneling microscopy (STM) images of few-layer graphene on the C face. At high resolution the graphene honeycomb lattice is revealed (Fig. 6.2a), and large-area STM scans [23] show that the top layer is extremely flat (Fig. 6.2b) on terraces of tens of micrometer size (Fig. 6.2c) in well-formed C-face graphene. Note the vertical scale of less than 0.1 nm in the 400-by-400 nm STM scan of Fig. 6.2b and the very low roughness (root mean square [RMS] < 50 pm) on the trace in Fig. 6.2d that extends over regions of different moiré patterns [24]. Extremely low roughness values are confirmed by X-ray diffraction (RMS < 5 pm) [25]. No grain boundaries have been seen in large-area STM scans (Fig. 6.2d), indicating the topmost graphene layer is continuous over the whole sample, which makes C-face epigraphene among the largest synthetic 2D crystals. The moiré patterns seen in Fig. 6.2d provide clear evidence that adjacently stacked graphene layers are rotated, creating a superstructure. Variations in the relative rotations of the layers change the moiré from place to place, with no change in the topographic height of the top layer (Fig. 6.2d). On a larger scale, smooth graphene pleats are observed, seen as bright lines in Fig. 6.2c (see also the scanning electron microscopy [SEM] image of Fig. 6.2e). The pleats (also referred to as puckers, ridges, wrinkles, folds, ripples, etc., in the literature) result from the larger thermal contraction of SiC compared to graphene when the films are cooled down from the growth temperature (around 1500°C–1600°C) to room temperature.

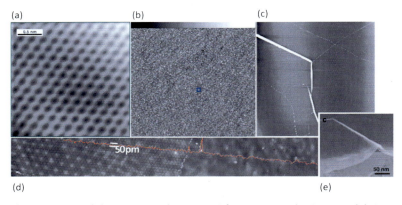

Figure 6.2 Multilayer epigraphene on C-face topography images. (a) STM image: 4 × 4 nm (adapted from Ref. [23]). (b) STM image: 400 × 400 nm. The blue square represents the size of the STM scan in (a) (adapted from Ref. [23]). (c) AFM scan: 40 μm × 40 μm. The white lines are graphene pleats. (d) 400 nm long STEM image across areas of different moiré patterns, showing that the top layer is flat (RMS < 50 pm) and continuous (Reprinted with permission from Ref. [35]. Copyright (2013) American Chemical Society). (e) Scanning electron microscopy image of a graphene pleat (Reprinted from Ref. [42], Copyright (2008), with permission from Elsevier).

The rotational layer stacking in C-face graphene is new and explicitly is not turbostratic (random stacking of small grains). In a CCS-grown C face, the stacking is determined by X-ray diffraction alternates 0 ± $\delta\theta$ degree–/30 ± $\delta\theta$ degree–oriented layers, with a variation of $\delta\theta$ within typically a few degrees [25]. Graphite (AB or rhombohedral) stacking in this case corresponds to stacking faults that have an occurrence of less than 15%–19% [25].

The properties of well-grown CCS C-face multilayered epitaxial graphene (MEG) are exceptional and have been amply discussed in the literature (see, for example, Ref. [1] for a review). It was soon realized that MEG is a model system to study graphene properties. This stems from a combination of factors: (i) excellent structural quality, with very few defects and continuous sheets, as shown by the absence of a D peak in Raman spectroscopy and by STM scans; (ii) flatness of the layers, which excludes strain-related gauge field effects and broadening effects in k-resolved surface spectroscopy measurements, such as angle-resolved photoemission spectroscopy (ARPES); (iii) extremely small interaction with the substrate and the environment, more specifically for the layers in the middle of the stack that are quasi neutral (charge density $n \approx 5 \times 10^9$ cm^{-2}, that is, at most 8 meV away from the Dirac point); and (iv) rotational stacking resulting in each layer in the stack having the electronic structure of monolayer graphene, explicitly not graphite.

Most prominently, the quasi-neutral layers show record high mobilities of 10^6 cm^2/V·s at room temperature, as determined by infrared magnetospectroscopy, with no sign of dependence on temperature [26]. Transport measurements that include contribution from the negatively doped layer at the interface consistently show high (but reduced) mobilities of the order of 15,000 to 25,000 cm^2/V·s [20, 27] and characteristics of graphene such as Berry's phase of π and weak antilocalization [1]. Spin transport is highly efficient in MEG, with record spin diffusion lengths up to 285 µm [27]. These layers show a textbook graphene electronic structure that allows spectroscopic studies (scanning tunneling spectroscopy [STS], ARPES, magnetospectroscopy, etc.) down to instrumental resolution [24, 26, 28]. Most notably, the fine structure of the Landau levels and their real space mapping were observed for the first time on a 2D system by STS measurement of the top C-face epigraphene layer. MEG is also very much sought for optical measurements due to the large area and homogeneity of the graphene layers residing on a transparent substrate. Moreover, since MEG is made of effectively decoupled graphene layers, the multiplicity enhances the signal without compromising the graphene characteristics, contrary to thin graphite. As an example of compelling results obtained from ultrafast optical spectroscopy studies in MEG, the electron dynamics probed reveal the details of electronic relaxation in graphene, as an interplay between efficient carrier–carrier and carrier–optical phonon scattering, and interlayer energy transfer effects [29].

6.2.2 Monolayer C Face

Monolayer graphene is more difficult to produce due to the rapid growth rates on the C face and the multiple nucleation sites. Usually isolated graphene patches up to tens of micrometers in size are produced by the CCS method [13, 22, 30] at a graphitization temperature around 1500°C. The graphene patches extend over multiple steps and show the characteristic pleat structure of C-face graphene (see Fig. 6.3a). With optimized growth conditions (typically around 1500°C for 10 min) larger patches and extended monolayer sheets can be produced as seen in the AFM image of Fig. 6.3a. An alternative method to CCS is to place a piece of graphite on the SiC chip, which provides some Si vapor confinement. This way narrow tapered stripes of monolayer graphene were produced [5].

The monolayer nature of the CCS-produced graphene is assessed with combined AFM imaging and Raman spectroscopy (narrow 2D peak), as shown in Fig. 6.3b. Raman spectroscopy is widely used to characterize graphene. The symmetrical and narrow graphene 2D peak observed in epigraphene provides an unambiguous signature of graphene monolayers. The D and G peaks are in the same energy range as the second-order SiC Raman peaks. A simple subtraction of the SiC contribution to the total spectra is generally performed, but for monolayer graphene a similar intensity of the SiC and graphene peaks results in poorly defined graphene spectra. The nonnegative matrix factorization (NMF) decomposition technique [31] allows the pure graphene and SiC spectra to be segregated from the raw data. This is done by decomposing a set of spectra, taken at different focal points of the exciting Raman laser light at and just below the surface, into a final number of spectral components. Details of the method are provided in Ref. [31]. A notable advantage of the technique is that the decomposition does not require prior knowledge of the components' spectral profile or even of the number of components needed (two in the case of graphene on SiC). The method applies to separate the contributions of stacked layers (such as graphene on SiC) or the contribution of patched films in inhomogeneous 2D maps. An example is given in Fig. 6.3b, where the Raman spectrum of single-layer graphene is extracted from the raw data. Significantly, the D peak is very small, attesting the good graphene quality.

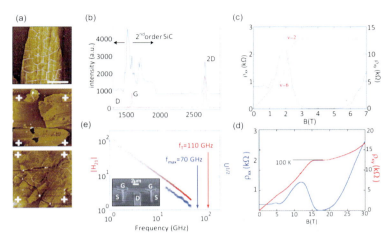

Figure 6.3 Monolayer graphene on C face. (a) AFM images of large monolayer patches draping over steps. (b) Raman spectroscopy of the monolayer C-face graphene in (a); the black trace is the raw data, and the red trace is the graphene spectrum once the SiC Raman peaks have been subtracted by the near-field microscopy (NFM) method. (c, d) Quantum Hall effect in monolayer graphene showing properties of the $\nu = 2$ plateau: (c) low field onset (3 tesla) for a sample of $\mu = 39,800$ cm^2/V·s and charge density $n = 0.19 \times 10^{12}$ cm^{-2} (From Ref. [13]. @ IOP Publishing. Reproduced with permission. All rights reserved.) and (d) high-temperature quantization ($n = 0.87 \times 10^{12}$ cm^{-2}; $\mu = 20,000$ cm^2/V·s (From Ref. [32]. @ IOP Publishing. Reproduced with permission. All rights reserved). (e) High-frequency transistor with a monolayer C-face graphene channel. The inset is the design of the dual-gate transistor; in the main panel the intercepts of $U^{1/2} = 1$ (slope 20 dB/dec) give a maximum oscillation frequency ($f_{max} = 70$G Hz) for a 100 nm gate length. U is Mason's unilateral gain (power gain). The current amplification H_{21} versus frequency gives a cutoff frequency $f_T = 110$ GHz (Reprinted with permission from Ref. [35]. Copyright (2013) American Chemical Society).

The epitaxial graphene monolayer is further unambiguously characterized from the observation of the quantum Hall effect (QHE) [13, 22] as shown in Fig. 6.3c,d for two samples of high-mobility $\mu = 39,800$ cm^2/V·s (20,000 cm^2/V·s) and charge density $n = 0.19 \times 10^{12}$ cm^{-2} (0.87 × 10^{12} cm^{-2}), respectively. Note that unlike the graphene layers in the middle of the stack in MEG that are quasi neutral, the graphene layer at the interface with SiC is naturally n doped with a carrier density of a few 10^{12} cm^{-2}. Positive counterdoping, either by exposure to the ambient or with a top electrostatic gate, is necessary to reduce the charge density close to the Dirac point [22]. The samples in Fig. 6.3 show two attractive characteristics of graphene for the resistance metrology standard based on the QHE, namely a robust Hall resistance quantization starting at a low magnetic field and still observed at a high temperature. The quantized $\rho_{xy} = 12.8$ k$\Omega = (2e^2/h)^{-1}$ Hall plateau in Fig. 6.3c has an extension of more than 4 tesla with an onset below 3 tesla, which is within the range of commercial room-temperature electromagnets. A very-well-defined plateau is observed at 100 K that is above the widely available liquid N_2 temperature (Fig. 6.3d) [32]. From a metrology perspective, note that the most accurate QHE plateau quantization that rivals the best GaAs 2DEGs QHE standards was measured in epigraphene [33, 34].

Having established a reliable production method of high-mobility monolayer graphene on the C face, the samples can be used for field-effect transistors (FETs) [35, 36]. For this the samples are patterned in a dual source/gate and common drain structure (Fig. 6.3e inset) that is designed for ultrahigh frequency measurements. By having contacts and T-shaped gates optimized to minimize access resistances and parasitic capacitances, C-face graphene FETs show a record maximum oscillation frequency $f_{max} = 70$ GHz [35]. The frequency f_{max} quantifies the practical upper bound for useful circuit operation. It was the first time that a power amplification was achieved at a frequency comparable to the current amplification (characterized by $f_T = 110$ GHz).

6.2.3   Monolayer Si Face

Epigraphene on the Si face is the most studied epigraphene form in the literature. Growth on the silicon face proceeds from the SiC steps. This adds an inherent difficulty to produce extended monolayer sheets because bilayers (or multiple layers) tend to form at step edges when full terrace graphene coverage is reached. As a matter of fact, the step edges act as electronic scattering centers, which has been related to a reduction of the carrier concentration on the sidewall of a step, to a n/p junction or the presence of a bandgap at the step edge or to the presence of bilayers/multilayers (see Ref. [1] and references therein). The latter have a detrimental effect on the homogeneity of the QHE. Better control of epigraphene growth on the Si face has been obtained by providing carefully balanced methane and H_2 [8].

The fact that growth starts at a sidewall can be turned into an advantage by arresting the growth on the sidewall of steps (either natural steps or etched steps in SiC—see in Section 3.2). Another consequence of graphene growth initiated at steps is that the quality of the SiC surface prior to growth is essential. Rough surfaces result in inhomogeneous and defective graphene layers, which can be rationalized by the presence of multiple graphene nucleation centers for surface decomposition. Surface flattening (removal of polishing scratches) can be obtained by SiC etching at high temperature (typically 1600°C) in a hydrogen environment (we use 3% H_2 in argon). Commercial SiC wafers are also available with surfaces treated by a chemical mechanical process, which provide good starting surfaces. For natural steps, another consideration is the width of the terraces, which depends on the step bunching for a given miscut. We have used pregrowth annealing in various conditions to help organize the step-terrace structure for a particular desired surface state. These include annealing at moderate temperature (1100°C–1200°C) in vacuum, high temperature in argon, and face-to-face SiC surface restructuring and step pinning under an evaporated refractory cap. In some cases it can be beneficial to grow first the buffer layer (the semiconducting graphene layer bound to SiC on the Si face) to help reduce the step displacement. An example of pregrowth SiC surface structuring is given in Fig. 6.4a, featuring large, atomically flat terraces with straight steps. The terraces are 20–40 μm in width and extend over hundreds of micrometers in length (see also Fig. 6.7e).

The capping technique involves evaporating an amorphous carbon grid onto bare SiC. Upon thermal treatment the SiC steps are pinned under the amorphous carbon and steps displacement is confined to within each amorphous carbon enclosure [37]. This results in step bunching at one end of the enclosure and step alignment along the amorphous carbon corral, as demonstrated in Fig. 6.4b,c. This technique demonstrates a significant improvement in SiC surface structuring by providing flat terraces not only of a large size (Fig. 6.4c, bottom) but most importantly at a predefined location. Note also that amorphous C is easily removable by plasma etching after annealing.

192 | Epitaxial Graphene on SiC

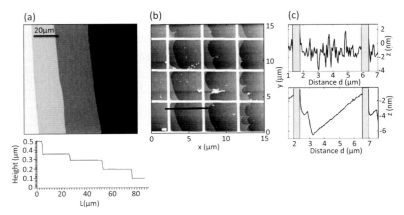

Figure 6.4 Structuring SiC. (a) AFM topographic image of 4H-SiC large terraces structured before growth and a step profile showing straight steps and flat terraces 20 to 40 μm wide. (b) SiC step pinning under an evaporated amorphous carbon grid. After annealing at 1350°C, the steps accumulate at one side of the grid in each enclosure, providing large terraces at locations defined by the grid (Reprinted from Ref. [37], with the permission of AIP Publishing). (c) AFM topographic profile within the grid (top) before and (bottom) after step-bunching annealing, showing that a large terrace has developed.

6.3 Selective Graphene Growth

6.3.1 Masking Techniques

Any graphene electronic application requires graphene patterning. The principle of structured growth stems from the realization that patterning a 2D graphene sheet by the usual oxygen plasma removal of carbon atoms is very destructive and leaves roughened edges and uncontrolled orientation and termination. It is, therefore, advantageous to grow graphene directly at preset locations. Figure 6.5a,b presents two examples where masking methods were used to grow graphene into shape with no need for further etching. Graphene growth can be either stopped under an AlN mask [14], slowed down under a Si-poor $Si_{3-x}N_4$ mask, or promoted under a rich $Si_{3+x}N_4$ mask [15]. In Fig. 6.5a, an 80 nm thick AlN film was evaporated and then patterned, revealing bare SiC in the shape of a multiprobe Hall bar. After annealing (20 min at 1420°C) multilayer graphene was grown on the 4H-SiC C face only on the exposed SiC, while no graphene grows under AlN, as shown by the Raman 2D map of Fig. 6.5a (bottom). In the case of the SiN mask, contrary to AlN, the mask vanishes during the heat treatment so that there is no need for further mask removal; a differential in graphene growth is revealed where the mask was present. The four panels of Fig. 6.5b show how the Si-rich $Si_{3+x}N_4$-patterned mask (a) was transferred into enhanced selective graphene growth, as shown in the optical image (b), the Raman 2D peak (c), and 2D peak/G peak maps (d) with submicron resolution. Conversely, graphene was used as a mask to prevent GaN growth in Fig. 6.5c. Here holes 75 nm in diameter are patterned in MEG (bright dots in the inset) and GaN crystals (triangles in the main panel) grow only on SiC in the holes. This structure is interesting with regard to integrating GaN LED with graphene.

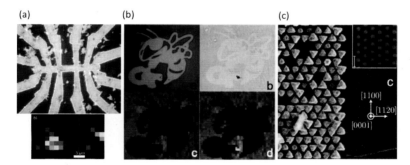

Figure 6.5 Selected growth. (a) Hall bar–shaped graphene multilayer (light gray) grown directly into shape using an AlN mask (dark) evaporated onto bare SiC. (Bottom) Raman map of the 2D graphene peak for the central area in (a), showing that graphene grows only in the unmasked areas (Reprinted from Ref. [14], with the permission of AIP Publishing). (b) Graphene growth under a Si-rich SiN mask (From Ref. [15]. @ IOP Publishing. Reproduced with permission. All rights reserved.): *(a)* Optical image of the SiN pattern, *(b)* optical image of subsequent MEG growth on SiC (graphene is light colored), *(c)* a Raman 2D map, and *(d)* a Raman 2D/G map, showing that graphene growth is promoted where the mask resided. The scale bar is 10 μm. (c) SEM images of 30 nm thick GaN grown on 4H-SiC using punctured graphene as a mask (Reprinted from Ref. [16], with the permission of AIP Publishing). No GaN grows on graphene, as exemplified on the right side of the image. Inset: SEM image of the epigraphene mask with 75 nm holes in it showing SiC (light dots). The scale bar in the inset is 200 nm.

6.3.2 Sidewall Facets

The concept of selective growth was applied to grow predefined nanoscale structures at desired locations without masking [12]. The technique relies on the fact that graphene grows faster on facets with orientation other than the (0001) basal plane of the Si face, as was readily observed by the growth of multilayer graphene at step edges (see earlier). Facets can be the natural steps arising from the miscut angle relative to the (0001) plane; step bunching by annealing in controlled condition produces arrays of steps and terraces with rather uniform width and height. More interestingly, trenches of various shapes can be etched in SiC [13], the sidewalls of which recrystallize into the crystallographic equilibrium facets of SiC—like (2-207) facets—upon annealing. Trenches can be etched, for instance, along the 4H-SiC-(-1-120) and 4H-SiC-(1-100) directions in SiC, providing graphene ribbons along the zigzag orientation and armchair orientation, respectively, owing to the epitaxial orientation of graphene on SiC (see Fig. 6.6a).

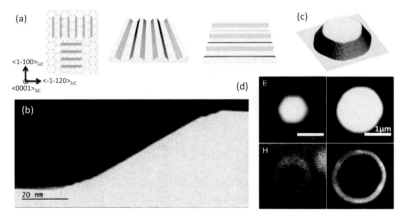

Figure 6.6 Sidewall growth. (a) Relative epigraphene to 4H-/6H-SiC orientation and direction of trenches to grow ribbons whose zigzag or armchair direction is, in principle, defined by the epitaxy of graphene on SiC. (b) Cross-sectional TEM image of a 40 nm high step covered with one-layer graphene (Reprinted with permission from Ref. [43]. Copyright (2014) American Chemical Society, available under the terms of the ACS AuthorChoice license). (c) 3D rendition of a facetted pillar after annealing. (d) AFM topographic (top) and electrostatic force microscopy (EFM) (bottom) image of annealed pillars, showing that graphene is confined to the sidewalls (bright ring in the EFM images) [3].

As for natural steps, graphene grows preferentially on these facets; under controlled growth conditions at a temperature of around 1600°C with the CCS method, graphene growth can be arrested on the sidewalls, producing monolayers on them. The ribbon width is then determined by the step height and the facet angle from the basal plane (about 27°). Because the steps in SiC are etched and annealed prior to graphene growth, the facets onto which graphene grows are smooth and atomically defined. This is best demonstrated in cross-sectional transmission electron microscopy (TEM). Figure 6.6b shows that a single graphene layer drapes over the (here armchair) step and merges into the buffer layer, which is tightly bound on the (0001) surface. Significantly, the graphene layer on the main facet is at a significantly larger distance from SiC than the buffer layer on the (0001) face.

Ribbons of any nominal orientation can in principle be produced. However, we have observed that straight steps may in some cases become rounded after annealing [17] and conversely etched pillars show faceting [3] along the armchair direction, as seen in Fig. 6.6c,d. Rounding and faceting are sensitive to a number of factors, such as step direction, growth condition (both temperature and time and possibly other factors, such as the type of heating), and the pinning of steps (e.g., under amorphous carbon pads, such as in Fig. 6.4b, or by defects). The SiC trench height is a relevant parameter since shallow trenches (≤10 nm) get washed out upon annealing and deep trenches tend to break into multiple facets, revealing multiple parallel ribbons. Uniform and well-formed graphene ribbons of various shapes are consistently realized on trenches 15–35 nm deep.

An important point of consideration is the ribbon integrity for transport measurements. Growth of secondary ribbons may occur if there are transverse substrate steps that cross the trenches. Graphene growth on these substrate steps produces "side arms" on the main ribbon that have been found to affect the transport [38]. An example is given on the AFM topographic and friction images of Fig. 6.7a,d. The substep features are clearly visible on the terraces of Fig. 6.7a (bright color) and give rise to ribbons seen as dark lines (low friction) in Fig. 6.7b. A single ribbon can be measured by selecting a segment with no side arms and etching away the surrounding, as was done in the low-temperature measurement in Ref. [17]. An alternative to get a very long, pristine ribbon is to choose a step/trench orientation parallel to the local miscut. That way, the terraces are flat (see Fig. 6.7d and the AFM trace in Fig. 6.7e) and very straight ribbons with no side arms can be formed, over hundreds of microns in length, as shown in Fig. 6.7d–f.

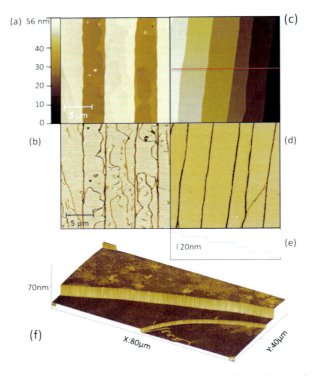

Figure 6.7 Perfecting sidewall ribbons. (a, b) Etched trenches with nonflat bottom and top terraces. (a) AFM topographic and (b) friction force microscopy (FFM) image (brown is a low-friction layer; yellow is the rippled, rougher buffer layer). Graphene grows on the main steps (vertical lines) as well as on the substeps on the terraces. (c, d) Straight isolated sidewall ribbons grown on bunched natural steps. (c) AFM and (d) FFM images for etched trenches with nonflat bottom and top terraces. (e) Topographic profile along the red line in (c). (f) AFM 3D image of a long isolated ribbon grown on a step.

Ribbons 30–50 nm wide show ballistic transport at room temperature [17]. This was unambiguously demonstrated in local four-point resistance measurements in UHV, showing a length-independent resistance from 1 to 17 microns at room temperature. Importantly, ballistic transport was observed not only for specific graphene orientations but also for chiral and curved ribbons. In an early experiment [39], a conductive AFM tip was brought into contact with a curved ribbon 35 nm wide and 1 µm long provided with a contact at both ends (Fig. 6.8a). The tip is scanned over the sample, and the tip-to-contact resistance is recorded. The resistance is minimum when the tip is in contact with the ribbons, as shown in the color plot of Fig. 6.8b. The minimum resistance R on each scan line is plotted as a function of the distance L to the contact in Fig. 6.8c. After subtracting the gold contact and line resistance, we find $R = R_{contact} + (\Delta R/\Delta L)L$, with $R_{contact} = 0.9 \, (h/e^2)$ and $\Delta R/\Delta L = 0.56 \, (h/e^2)/\mu m$. The contact resistance indicates there is only one conducting channel; in these conditions [40] the slope $\Delta R/\Delta L = (h/e^2)(L/\lambda)$ gives a mean free path $\lambda = 1.8$ µm. This value is longer than the ribbon length, indicating ballistic transport. This result is remarkable because the ribbon is exposed to ambient conditions, which in general increases significantly the resistance in contrast to protected ribbons in UHV or encapsulated under a dielectric (Al$_2$O$_3$ like in Ref. [17]). We believe that scanning the tip over the ribbon cleanses the ribbon by sweeping away the contaminants. We have often observed accumulation of particles on the rougher buffer layer next to ribbons.

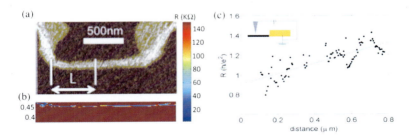

Figure 6.8 Electrical transport in sidewall ribbons. (a) EFM image of an isolated epigraphene ribbon connected to two large graphene pads. (b) Tip to sample resistance color map showing that conduction is along the ribbon only (the resistance scale is on the right). (c) Plot of the resistance as a function of length along the ribbon in ambient conditions.

6.4 Large-Scale Integration

By design, multiple ribbons can be produced all at once. Arrays of tens of thousands of ribbons have been grown on the sidewall of trenches patterned by photolithography [12]. An example of integration of 10,000 FETs/cm^2 is presented in the optical image of Fig. 6.9, where an aluminum gate (G) lies on a ribbon in between the Pd/Au source and drain contacts (S–D).

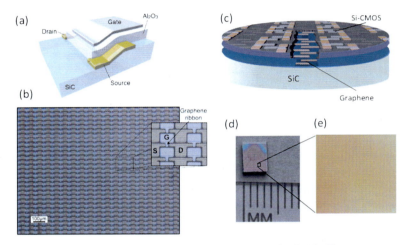

Figure 6.9 Large-scale integration. (a) Schematics of a field-effect transistor (FET) with a sidewall ribbon as the channel connected to source S and drain D with a top gate G. (b) Integration of 10,000 FETs/cm^2 (optical image) with parallel ribbons connected between Pd/Au contacts and provided with an aluminum gate (G) on Al$_2$O$_3$ [12]. (c) Principle of epigraphene on SiC-to-Si wafer bonding, with a top Si wafer ready for CMOS technology and an epigraphene circuit underneath. Both are interconnected by metal vias through the thin Si layer. (d, e) Realization of the epigraphene on SiC-to-Si wafer bonding. (d) Optical image of the biwafer chip from the Si side. The purple color corresponds to the strongly bonded regions. (e) Optical image from the transparent SiC side, showing an intact trench array supporting the sidewall ribbons (From Ref. [41]. @ IOP Publishing. Reproduced with permission. All rights reserved).

Mainstream electronics and very large-scale integration are based on silicon; it is, therefore, important to devise schemes to integrate graphene with Si wafers. The route we proposed [41] is based on the standard industry methods using wafer bonding and the smart-cut technique. The idea here is to transfer a Si film from a

Si wafer onto a graphitized SiC wafer. Figure 6.9c shows the principle of the design. The top-bounded Si wafer is ready for device patterning, and the graphene layer beneath is accessed through metallic vias managed through the submicron-thick Si wafer. This 3D integration realizes the interconnection of the SiC-supported graphene platform and the Si-based electronic wafer while preserving the integrity of graphene. Details have been published elsewhere; the main steps consist of the fabrication of graphene structures (either from patterned epitaxial graphene or from template growth on SiC sidewalls), evaporation of an alumina film that serves as a bonding layer between the SiC and the oxidized Si wafer, wafer bonding, and finally splitting of the Si wafer (smart cut) that leaves a thin crystalline Si layer on top. The main advantage of the process is that the graphene growth temperature (above 1500°C) is not limited by the presence of silicon (melting point 1414°C) or the limited SiC quality, like in SiC epilayers on Si; the Si wafer is fully accessible, and the stacking of wafers increases the areal density inspired by the 3D-stacked layers' very-large-scale-integration technology.

Figure 6.9d,e shows the top Si film bonded on a SiC chip. The purple color corresponds to the strongly bonded regions. The zoomout optical image of Fig. 6.9e shows an array of ribbons grown on parallel sidewall trenches, as seen from the SiC (transparent) side of the bonded wafers. Note that the ribbon structure is intact after bonding.

6.5   Conclusion

We have reviewed the main characteristics of epitaxial graphene grown on SiC by the CCS method and selective graphene growth techniques. Most recent results on structured growth demonstrate that graphene nanoribbons can be grown on the sidewall of either natural SiC steps or trenches etched in SiC. Ribbons 20–40 nm wide and more than 100 µm long can be very well confined onto the sidewall and well isolated (no side arms) from other graphene structures. Schemes for large-scale integration of epitaxial graphene ribbons with Si wafer were also presented.

Acknowledgments

Financial support is acknowledged from the Air Force Office of Scientific Research (AFOSR) and the National Scientific Foundation (NSF) under grants FA9550-13-1-0217 and 1506006, respectively. Additional support is provided by the Partner University Fund from the French Embassy. CB acknowledges partial funding from the EU Graphene Flagship program.

References

1. Berger, C., Conrad, E. H., de Heer, W. A., *Epigraphene*, Chiaradia, P., and Chiarotti, G. (eds.), Landolt-Borstein (2016).

2. Berger, C., Song, Z. M., Li, T. B., Li, X. B., Ogbazghi, A. Y., Feng, R., Dai, Z. T., Marchenkov, A. N., Conrad, E. H., First, P. N., de Heer, W. A., Ultrathin epitaxial graphite: 2D electron gas properties and a route toward graphene-based nanoelectronics, *J. Phys. Chem. B*, **108**, 19912–19916 (2004).

3. de Heer, W. A., Berger, C., Ruan, M., Sprinkle, M., Li, X., Hu, Y., Zhang, B., Hankinson, J., and Conrad, E. H., Large area and structured epitaxial graphene produced by confinement controlled sublimation of silicon carbide, *Proc. Natl. Acad. Sci. U S A*, **108**, 16900–16905 (2011).

4. Emtsev, K. V., Bostwick, A., Horn, K., Jobst, J., Kellogg, G. L., Ley, L., McChesney, J. L., Ohta, T., Reshanov, S. A., Rohrl, J., Rotenberg, E., Schmid, A. K., Waldmann, D., Weber, H. B., and Seyller, T., Towards wafer-size graphene layers by atmospheric pressure graphitization of silicon carbide, *Nat. Mater.*, **8**, 203–207 (2009).

5. Camara, N., Huntzinger, J. R., Rius, G., Tiberj, A., Mestres, N., Perez-Murano, F., Godignon, P., and Camassel, J., Anisotropic growth of long isolated graphene ribbons on the C face of graphite-capped 6H-SiC, *Phys. Rev. B*, **80**, 125410 (2009).

6. Yakes, M. K., Gunlycke, D., Tedesco, J. L., Campbell, P. M., Myers-Ward, R. L., Eddy, C. R., Gaskill, D. K., Sheehan, P. E., and Laracuente, A. R., Conductance anisotropy in epitaxial graphene sheets generated by substrate interactions, *Nano Lett.*, **10**, 1559–1562 (2010).

7. Robinson, J., Weng, X. J., Trumbull, K., Cavalero, R., Wetherington, M., Frantz, E., LaBella, M., Hughes, Z., Fanton, M., and Snyder, D., Nucleation of epitaxial graphene on SiC(0001), *ACS Nano*, **4**, 153–158 (2010).

8. Michon, A., Vezian, S., Roudon, E., Lefebvre, D., Zielinski, M., Chassagne, T., and Portail, M., Effects of pressure, temperature, and hydrogen during graphene growth on SiC(0001) using propane-hydrogen chemical vapor deposition, *J. Appl. Phys.*, **113**, 203501 (2013).

9. Suemitsu, M., Jiao, S., Fukidome, H., Tateno, Y., Makabe, I., and Nakabayashi, T., Epitaxial graphene formation on 3C-SiC/Si thin films, *J. Phys. D*, **47**, 094016 (2014).

10. Yakimova, R., Iakimov, T., Yazdi, G. R., Bouhafs, C., Eriksson, J., Zakharov, A., Boosalis, A., Schubert, M., and Darakchieva, V., Morphological and electronic properties of epitaxial graphene on SiC, *Physica B*, **439**, 54–59 (2014).

11. Robinson, J. A., Hollander, M., LaBella, M., Trumbull, K. A., Cavalero, R., and Snyder, D. W., Epitaxial graphene transistors: enhancing performance via hydrogen intercalation, *Nano Lett.*, **11**, 3875–3880 (2011).

12. Sprinkle, M., Ruan, M., Hu, Y., Hankinson, J., Rubio-Roy, M., Zhang, B., Wu, X., Berger, C., and de Heer, W. A., Scalable templated growth of graphene nanoribbons on SiC, *Nat. Nanotechnol.*, **5**, 727–731 (2010).

13. Hu, Y., Ruan, M., Guo, Z. L., Dong, R., Palmer, J., Hankinson, J., Berger, C., and de Heer, W. A., Structured epitaxial graphene: growth and properties, *J. Phys. D*, **45**, 154010 (2012).

14. Rubio-Roy, M., Zaman, F., Hu, Y. K., Berger, C., Moseley, M. W., Meindl, J. D., and de Heer, W. A., Structured epitaxial graphene growth on SiC by selective graphitization using a patterned AlN cap, *Appl. Phys. Lett.*, **96**, 082112 (2010).

15. Puybaret, R., Hankinson, J., Palmer, J., Bouvier, C., Ougazzaden, A., Voss, P. L., Berger, C., and de Heer, W. A., Scalable control of graphene growth on 4H-SiC C-face using decomposing silicon nitride masks, *J. Phys. D*, **48**, 152001 (2015).

16. Puybaret, R., Patriarche, G., Jordan, M. B., Sundaram, S., El Gmili, Y., Salvestrini, J. P., Voss, P. L., de Heer, W. A., Berger, C., and Ougazzaden, A., Nanoselective area growth of GaN by metalorganic vapor phase epitaxy on 4H-SiC using epitaxial graphene as a mask, *Appl. Phys. Lett.*, **108**, 103105 (2016).

17. Baringhaus, J., Ruan, M., Edler, F., Tejeda, A., Sicot, M., Taleb-Ibrahimi, A., Li, A. P., Jiang, Z. G., Conrad, E. H., Berger, C., Tegenkamp, C., and de Heer, W. A., Exceptional ballistic transport in epitaxial graphene nanoribbons, *Nature*, **506**, 349–354 (2014).

18. Van Bommel, A. J., Crobeen, J. E., and Van Tooren, A., LEED and Auger electron observations of the SiC(0001) surface, *Surf. Sci.*, **48**, 463–472 (1975).

19. Palmer, J., *Pre-Growth Structures for Nanoelectronics of EG on SiC*, PhD dissertation, Georgia Institute of Technology (2014), https://smartech.gatech.edu/handle/1853/54293.

20. Berger, C., Song, Z. M., Li, X. B., Wu, X. S., Brown, N., Naud, C., Mayou, D., Li, T. B., Hass, J., Marchenkov, A. N., Conrad, E. H., First, P. N., and de Heer, W. A., Electronic confinement and coherence in patterned epitaxial graphene, *Science*, **312**, 1191–1196 (2006).

21. Maysonnave, J., Huppert, S., Wang, F., Maero, S., Berger, C., de Heer, W. A., Norris, T. B., de Vaulchier, L. A., Dhillon, S., Tignon, J., Ferreira, R., and Mangeney, J., Terahertz generation by dynamical photon drag effect in graphene excited by femtosecond optical pulses, *Nano Lett.*, **14**, 5797–5802 (2014).

22. Wu, X. S., Hu, Y. K., Ruan, M., Madiomanana, N. K., Hankinson, J., Sprinkle, M., Berger, C., and de Heer, W. A., Half integer quantum Hall effect in high mobility single layer epitaxial graphene, *Appl. Phys. Lett.*, **95**, 223108 (2009).

23. Haas, J., *Structural Characterization of Epitaxial Graphene on Silicon Carbide*, PhD thesis, Georgia Institute of Technology (2008).

24. Miller, D. L., Kubista, K. D., Rutter, G. M., Ruan, M., de Heer, W. A., First, P. N., and Stroscio, J. A., Observing the quantization of zero mass carriers in graphene, *Science*, **324**, 924–927 (2009).

25. Hass, J., de Heer, W. A., and Conrad, E. H., The growth and morphology of epitaxial multilayer graphene, *J. Phys.: Condens. Matter*, **20**, 323202 (2008).

26. Orlita, M., Faugeras, C., Grill, R., Wysmolek, A., Strupinski, W., Berger, C., de Heer, W. A., Martinez, G., and Potemski, M., Carrier scattering from dynamical magnetoconductivity in quasineutral epitaxial graphene, *Phys. Rev. Lett.*, **107**, 216603 (2011).

27. Dlubak, B., Martin, M. B., Deranlot, C., Servet, B., Xavier, S., Mattana, R., Sprinkle, M., Berger, C., de Heer, W. A., Petroff, F., Anane, A., Seneor, P., and Fert, A., Highly efficient spin transport in epitaxial graphene on SiC, *Nat. Phys.*, **8**, 557–561 (2012).

28. Sprinkle, M., Siegel, D., Hu, Y., Hicks, J., Tejeda, A., Taleb-Ibrahimi, A., Le Fevre, P., Bertran, F., Vizzini, S., Enriquez, H., Chiang, S., Soukiassian, P., Berger, C., de Heer, W. A., Lanzara, A., and Conrad, E. H., First direct observation of a nearly ideal graphene band structure, *Phys. Rev. Lett.*, **103**, 226803 (2009).

29. Mihnev, M. T., Kadi, F., Divin, C. J., Winzer, T., Lee, S., Liu, C.-H., Zhong, Z., Wang, X., Ruoff, R. S., Berger, C., de Heer, W. A., Malic, E., Knorr, A., and Norris, T.B., Microscopic origins of the terahertz carrier relaxation and cooling dynamics in graphene, *Nat. Commun.*, **7**, 11617 (2016).

30. Zhang, R., Dong, Y. L., Kong, W. J., Han, W. P., Tan, P. H., Liao, Z. M., Wu, X. S., and Yu, D. P., Growth of large domain epitaxial graphene on the C-face of SiC, *J. Appl. Phys.*, **112**, 104307 (2012).

31. Kunc, J., Hu, Y., Palmer, J., Berger, C., and de Heer, W. A., A method to extract pure Raman spectrum of epitaxial graphene on SiC, *Appl. Phys. Lett.*, **103**, 201911 (2013).

32. de Heer, W. A., Berger, C., Wu, X., Hu, Y., Ruan, M., Stroscio, J., First, P., Haddon, R., Piot, B., Faugeras, C., and Potemski, M., Epitaxial graphene electronic structure and transport, *J. Phys. D*, **43**, 374007 (2010).

33. Tzalenchuk, A., Lara-Avila, S., Kalaboukhov, A., Paolillo, S., Syvajarvi, M., Yakimova, R., Kazakova, O., Janssen, T. J. B. M., Fal'ko, V., and Kubatkin, S., Towards a quantum resistance standard based on epitaxial graphene, *Nat. Nanotechnol.*, **5**, 186–189 (2010).

34. Lafont, F., Ribeiro-Palau, R., Kazazis, D., Michon, A., Couturaud, O., Consejo, C., Chassagne, T., Zielinski, M., Portail, M., Jouault, B., Schopfer, F., and Poirier, W., Quantum Hall resistance standards from graphene grown by chemical vapour deposition on silicon carbide, *Nat. Commun.*, **6**, 6806 (2015).

35. Guo, Z. L., Dong, R., Chakraborty, P. S., Lourenco, N., Palmer, J., Hu, Y. K., Ruan, M., Hankinson, J., Kunc, J., Cressler, J. D., Berger, C., and de Heer, W. A., Record maximum oscillation frequency in C-face epitaxial graphene transistors, *Nano Lett.*, **13**, 942–947 (2013).

36. Lin, Y. M., Dimitrakopoulos, C., Jenkins, K. A., Farmer, D. B., Chiu, H. Y., Grill, A., and Avouris, P., 100-GHz transistors from wafer-scale epitaxial graphene, *Science*, **327**, 662–662 (2010).

37. Palmer, J., Kunc, J., Hu, Y. K., Hankinson, J., Guo, Z. L., Berger, C., and de Heer, W. A., Controlled epitaxial graphene growth within removable amorphous carbon corrals, *Appl. Phys. Lett.*, **105**, 023106 (2014).

38. Baringhaus, J., Aprojanz, J., Wiegand, J., Laube, D., Halbauer, M., Hubner, J., Oestreich, M., and Tegenkamp, C., Growth and characterization of sidewall graphene nanoribbons, *Appl. Phys. Lett.*, **106**, 043109 (2015).

39. Sidorov, A., Ruan, M., Berger, C., Jiang, Z., and de Heer, W. A., unpublished (2010).

40. Datta, S., *Electronic Transport in Mesoscopic Systems*, Cambridge University Press, Cambridge (1995).

41. Dong, R., Guo, Z. L., Palmer, J., Hu, Y. K., Ruan, M., Hankinson, J., Kunc, J., Bhattacharya, S. K., Berger, C., and de Heer, W. A., Wafer bonding solution to epitaxial graphene-silicon integration, *J. Phys. D*, **47**, 094001 (2014).

42. Cambaz, Z. G., Yushin, G., Osswald, S., Mochalin, V., and Goyotsi, Y., Noncatalytic synthesis of carbon nanotubes, graphene and graphite on SiC, *Carbon*, **46**, 841–849 (2008).

43. Palacio, I., Celis, A., Nair, M. N., Gloter, A., Zobelli, A., Sicot, M., Malterre, D., Nevius, M. S., de Heer, W. A., Berger, C., Conrad, E. H., Taleb-Ibrahimi, A., and Tejeda, A., Atomic structure of epitaxial graphene sidewall nanoribbons: flat graphene, miniribbons, and the confinement gap, *Nano Lett.*, **15**, 182–189 (2014).

Index

2D bands, 55
π-bands, 50–52, 58, 96–97, 143, 145–46, 149–50, 154, 158
  linear, 143
  linear dispersing, 146
σ-band, 96,

ABA-stacked trilayer graphene, 52
ABA-stacked trilayers, 151–52
ABA stacking, 81, 151, 156, 170
ABA trilayer, 53
ABC trilayers, 151
absorption, 117, 183
  near-edge X-ray, 38
  optical, 116
activation energy, 124, 126–27, 130–32
adhesion, low, 2, 13
adhesion energy, 4, 13, 18
adlayer, 83, 159
adsorption, 161, 185
AFM, *see* atomic force microscopy
amorphous carbon, 7, 191, 195
amorphous carbon grid, 191–92
angle-resolved photoemission, 50
angle-resolved photoemission spectroscopy (ARPES), 30, 32, 46, 50–53, 56–58, 96, 98, 145–46, 148–52, 154, 157–58, 160, 167, 170, 187
angular dependence, 39, 57, 168
annealing, 9–10, 12–14, 78–80, 82–83, 86–87, 89–90, 92–93, 110–14, 120–21, 146, 148–50, 154–55, 158–59, 161–63, 191–95
annealing temperature, 35–36, 82, 87, 111, 114–15, 121, 123, 161–62

annealing time, 116–18, 121, 131
antiphase boundary (APB), 11
antiphase domain (APD), 42, 57, 80
antilocalization, weak, 187
AP, see atmospheric pressure
APB, *see* antiphase boundary
APD, *see* antiphase domain
APD boundaries, 42–44, 48, 55, 57–58
argon atmosphere, 31, 34, 55
argon pressure, 116–17
armchair direction, 194–95
armchair structure, 62
ARPES, *see* angle-resolved photoemission spectroscopy
ARPES spectrum, 97, 150, 161–62
Arrhenius formula, 114
Arrhenius function, 115
Arrhenius plot, 124, 126, 130, 132
asymmetry, 58, 154
asymptotic conditions, 128
asymptotic value, 122–23, 128
atmospheric pressure (AP), 144–45, 147–48, 150
atom diffusion, 128, 132
atom evaporation, 117
atomic force microscopy (AFM), 96–97, 162, 185–86, 188–89, 192, 194–96
atomic Intercalation, 141
atomic structure, 31–32, 35, 38, 46, 59, 62, 92, 95
azimuthal directions, 99

back-gate field effect transistors, 11
band folding, 168

bandgap, 31, 42, 62, 148–52, 154–55, 168, 170, 190
bands, 52, 55, 57, 98–99, 151, 155, 168
 2D, 150, 154
 experimental, 155
 linearly dispersing, 151
 low-energy, 151
 parabolic, 150, 154
 quadratically dispersing, 151
 theoretical, 154–55
 tunable, 151
 valence, 99
band structure, 30, 52, 84, 149, 151–52, 154, 160, 163
 electronic, 145, 152, 155, 168
 low-energy, 151
band velocity, 157, 168
BE, *see* binding energy
Bernal stacking, 33, 93, 151
BF, *see* bright field
bias, 42, 91
bias voltages, 11, 42, 50–51, 60
bilayer graphene (BLG), 13, 46, 55, 80, 88, 95–96, 110, 142, 144, 148–50, 152–54, 169–70
 as-grown, 150, 153
 quasi-freestanding, 148, 150
binding energy (BE), 33, 36, 40, 46, 111, 166
BLG, *see* bilayer graphene
BN, *see* boron nitride
bonding, covalent, 144
bonding layer, 199
boundaries, 42, 50, 54, 58, 62–63, 131, 135, 169
 antiphase, 11
 grain, 185
 mBZ, 168–69
 twin, 89
boron nitride (BN), 3, 29
boundary conditions, 90, 130, 133
bright field (BF), 57

Brillouin zone (BZ), 50–52, 58, 167–68
buffer layer, 7, 9, 32–33, 83–84, 86, 93–94, 112–13, 115, 117, 119, 121, 156, 161, 191, 195–97
bulk component, 36, 39
bulk crystals, 33, 81, 114
BZ, *see* Brillouin zone

calibrated hole, 183–84
capping technique, 191
carbon atomic chains, 53–54
carbon atoms, 4, 7, 29, 32–33, 36, 38, 54, 79, 142–43, 156, 166, 192
 dissolved, 4
 excessive, 37
 extra, 37, 42
 sp^2-bonded, 11
 sp^2-hybridized, 28
carbon–carbon bond, 36, 48–49, 53
carbon chains, 53
carbon ion implantation, 4–5, 7
carbonization, 28
carbon layer, 85, 143, 146, 151, 160
carbon nanotube (CNT), 2
carrier–carrier scattering, efficient, 187
carrier density, 161, 189
carrier mobility, 2, 145, 158
 high, 9, 12, 31, 158
 reduced, 158
 ultrahigh, 1
carrier–optical phonon scattering, 187
carriers, charge, 8, 29, 59, 62
CCS, *see* confinement-controlled sublimation
C-face epigraphene, 185
C-face graphene, 185–86, 188
C-face graphene FETs, 190
C-face topography images, 186

Index | 207

charge neutrality, 50, 146, 150, 155, 158
charge transfer, 33, 84, 151, 168
charge transport gap, 60, 62–63
chemical dopant, 149
chemical exfoliation, 28
chemical vapor deposition (CVD), 2, 29, 31, 116, 134, 136
chip, biwafer, 198
clusters, 129
    stable, 129
    tetrahedral adatom, 83
CMOS, *see* complementary-metal-oxide semiconductor
CNT, *see* carbon nanotube
coevaporation, 120
coincidence superlattice, 163, 166
complementary metal-oxide semiconductor (CMOS), 28, 198
concentration, carrier, 161, 169, 190
conduction band, 147, 155
confinement-controlled sublimation (CCS), 8, 79, 181–86, 188, 195, 199
constant
    normalization, 129
    Planck's, 60
contacts
    drain, 198
    source, 198
    transparent electrical, 27
continuity, 11, 42–43, 48, 63, 80, 95
contours, Fermi, 167
copper, single-crystal, 165
core-level spectra, 12, 33–35, 38–39, 45, 115, 150
corrugation, 147, 151
    peak-to-peak, 147
covalent bonds, 84, 147
crystal lattice, 31, 42, 48, 53, 59
crystals, 79, 157

    2D, 28, 79, 160, 185,
    HOPG, 28
    Si, 157
    SiC, 9, 33, 36, 40, 42, 48, 53, 59, 184
cubic dispersion, 151
cubic lattice, 32
Cu intercalation, 165
current amplification, 189–90
CVD, *see* chemical vapor deposition

dangling bonds, 146, 155
dark field (DF), 44
decomposition, 127, 143, 183, 188
deconvolution, 40
decoupling, 9, 144, 146, 163
defect density, 13, 80
defect-free graphene, 96
defects, 41–42, 44, 48, 51, 54–55, 61, 63, 79–80, 89, 92, 94, 118, 120, 123, 125
    heptagon-pentagon, 110, 118
    intrinsic, 48
degradation, 31
deintercalation, complete, 148
density, 54, 89, 133, 170
    areal, 199
    charge, 187, 189
    high, 52, 143, 163
    lower atom, 12
    silicon vapor, 79
density functional theory (DFT), 89–90, 99
density of defects, 11, 55, 57
density profile, 133, 135
deposition rate, 7, 80
desorption, 95, 110, 118–19
device fabrication, 2, 13, 29
    electronic, 15
    large-scale, 13
devices, 11, 60
    advanced biomedical, 157
    bottom-gated EG, 158
    electronic, 27

functional, 13
gap-tunable, 156
graphene-based, 9, 15
hard, 8
light-emitting, 8
microelectronic, 29
nanoelectronic, 29
nanogap contact, 60
nanointegrated, 2
nanoscale, 109
photonic, 27
DF, *see* dark field
DF LEEM images, 44–45, 47
DFT, *see* density functional theory
dielectric insulators, 2
dielectrics, 197
diffusion, 79, 95, 110, 117–18, 126, 130–31, 133
diffusion barrier, 119, 139
diffusion coefficient, 120, 125, 130, 135
diffusion equation, 120, 133, 135
dipole, additional, 151
Dirac cone, 30, 50, 52, 97–99, 168–69
Dirac dispersion, 84
Dirac electrons, 158
Dirac energy, 154
Dirac fermions, 160, 166
Dirac nature, 8
Dirac points, 50, 63
direct growth, 3–4, 15, 24
direct visualization, 152, 155
dispersion, 50–52, 58, 143, 146, 149–50, 154, 168
dispersion
ARPES, 52
electronic, 151, 155
distortions, 48, 55, 92
carbon–carbon bond, 63
random, 53
distribution
bond length, 48
spatial, 55

domain boundaries, 42, 48, 51, 53, 55, 59
<110>-directed, 47
antiphase, 63
domain network, 50, 53
domains, 9, 42, 46, 48, 50–51, 58–60, 88–89, 94, 116, 153, 155, 157
antiphase, 42, 57, 80
continuous, 99, 110
large-scale, 145
nanometer-scale, 58
doping, 84, 145, 149–50, 158–59, 161
charge, 52
chemical, 78
extrinsic, 160
hole, 149
molecular, 149, 173
dual-gate transistor, 189

EDC, *see* energy distribution curve
EFM, *see* electrostatic force microscopy
electric dipole, 148
electron beam lithography, 60
electron doping, 145, 159
electron energy, 45, 58, 87, 153
electron–hole pairs, 158
electronic devices, 2, 8, 31, 145
SiC-based high-temperature, 31
electronic relaxation, 187
electronic structure, 10, 28, 31, 41, 50, 52, 56, 187
electron microscopy, 46
electron mobility, 84
electron optics, 87
electron reflectivity, 45, 57–58, 153–54
electrostatic force microscopy (EFM), 194, 197
energy bandgap, 59, 155
energy dispersion, 168–69

energy distribution curve (EDC), 97, 99

epigraphene, 182–83, 188, 190, 193, 198

epilayers, 31–33, 60, 81–82, 85, 95, 199

epitaxial graphene, 9–11, 32, 78, 80, 84–87, 94, 96, 99, 115–16, 120, 141, 143, 169, 181–83, 199
  Bernal stacked, 86, 110, 123
  growth mode, 110
  large-area, 94
  monolayer, 181
  multilayered, 8, 111, 186
  patterned, 199

epitaxial graphene domains, 94

epitaxial graphene films, 111, 116, 133

epitaxial graphene growth, 81, 86, 110, 114, 121

epitaxial graphene layers, 12, 32, 56, 78–79, 84, 87–88

epitaxial graphene monolayer, 189

epitaxial graphene ribbons, 199

epitaxial growth, 8, 10, 78, 81, 124

epitaxial layer, 78, 165

epitaxial orientation, 194

epitaxial registry, 165

epitaxy, 78, 86, 115

equilibrium, 79, 110, 116, 183–84

etching, 7, 191–92, 195

evaporation, 112, 116, 120, 124, 126–27, 131, 199

exfoliated graphene, 2, 12, 48, 55

fabrication, 15, 29, 34–35, 37, 55, 84, 145, 156, 158, 199
  industrial-scale, 80
  large-scale, 2, 15
  wafer-level, 14
  wafer-scale, 13

faceting, 195

facets, 194–95

crystallographic equilibrium, 194

fast Fourier transform (FFT), 46–48, 93–94, 156

Fermi level, 50, 52, 57, 63, 98, 145, 150, 154, 158, 168

fermions, massive, 170

Fermi surface (FS), 167–69

Fermi velocity, 60

FET, *see* field-effect transistor

few-layer graphene, 4, 12, 14, 28–31, 35, 37–39, 41, 63, 93, 98, 152, 185
  nanostructured, 46, 53
  uniform, 63

FFM, *see* friction force microscopy

FFT, *see* fast Fourier transform

field-effect transistor (FET), 3, 149, 190, 198

field of view (FOV), 88, 153, 161–2

film closure, 129

films, thin, 32, 55, 63, 85–86, 114

first-principles calculations, 160

flakes, 2, 28, 145

flash-heating, 35

flux, 82–83, 86–87, 110–12, 120, 125, 128, 132, 134

Fourier components, 94

Fourier transform, back, 94

FOV, *see* field of view

freestanding graphene, 44, 48–49, 146

frequency, 189–90
  cutoff, 158
  maximum oscillation, 189

friction, low, 195

friction force microscopy (FFM), 196

FS, *see* Fermi surface

full-width at half-maximum (FWHM), 4, 39, 41, 52–53, 55–56

functionalization, 78, 149

molecular, 151
FWHM, *see* full-width at half-maximum

gap resistance, 50–51
gas-source molecular beam epitaxy (GSMBE), 10
Gaussian function, 97
Gaussian maxima, 98
Gaussian peak, 99
Ge-intercalated graphene, 162
Ge intercalation, 162
GOS, *see* graphene on silicon
GOSFET, *see* graphene-on-silicon field-effect transistor
graphene, 1–14, 27–30, 32–33, 37–47, 49–55, 77–84, 88–99, 109–14, 116–22, 124–28, 156–61, 163–70, 182–84, 186–88, 192–96
  alloy-mediated, 13
  as-grown, 146
  charge-neutral, 160
  decoupled, 146
  electronic properties of, 83, 86, 114, 157
  epitaxy of, 86, 114, 194
  highest-quality, 48
  high-quality, 2, 4, 10, 12, 15–16, 55, 79, 157
  homogeneous, 147, 157
  hydrogen-intercalated, 138, 150
  large-area, 7, 17
  micrometer-scale, 42
  mixed-phase, 162
  neutral, 166
  n-phase, 162
  one-layer, 194
  p-phase, 162
  punctured, 193
  quasi-freestanding, 63, 142, 147, 157

  semimetallic, 54, 63

  suspended, 158
  trilayer, 54
  unperturbed, 167
  wafer-scale, 55
  zero-layer, 144
graphene band structure, 160, 168
graphene Brillouin zone, 30, 99, 145–46, 149–50, 154
graphene buffer layer, 185
graphene corrugation, 73, 179
graphene coverage, 37, 44, 48, 63, 190
graphene devices, 2, 8
graphene domains, 9, 44, 46, 48, 53–54
graphene fabrication, 28, 31
graphene film, 9, 11–12, 30, 110, 116, 154
graphene flakes, 2, 15, 143, 149
graphene formation, 40, 85, 89, 96, 110, 124, 127
graphene growth, 3, 9–10, 78–80, 83–84, 96, 99, 109–11, 113–14, 116–17, 119–21, 127, 136, 144, 191–93, 195
  continuous, 118
  direct, 3, 10, 15, 78
  enhanced selective, 193
  few-layer, 40, 53, 66
  high-quality, 15
  metal-mediated, 10, 15
  solid-source-based MBE, 7
graphene hexagons, 166
graphene honeycomb lattice, 32, 36, 142, 185
graphene islands, 131, 136
graphene lattices, 46, 48, 53–54, 60, 92, 145, 147, 164–65
graphene layers, 5–7, 9–11, 53–54, 79–80, 83–84, 92–93, 95–98, 114–19, 123–24, 142–43, 147, 159, 165–66, 183, 185
  atomic structure, 95

continuous, 80, 94
decoupled, 159, 187
defective, 191
high-crystalline-quality, 183
homogeneous, 55
nonuniform, 88
planar, 29
semiconducting, 191
graphene monolayers, 30, 49, 57, 143, 188
graphene multilayers, 193
graphene nanodomains, 37
graphene nanoribbons, 59, 170, 199
  self-aligned, 54, 59
graphene nanostructures, 15, 181–82
graphene nuclei, 127–28, 132–33
graphene on silicon (GOS), 11, 32–33
graphene-on-silicon field-effect transistor (GOSFET), 11–12
graphene overlayer, 28–29, 31–37, 39–44, 48, 53–55, 58, 63, 93, 114
graphene patches, 188
graphene patterns, 30, 192
graphene π-bands, 168
graphene pleats, 186
graphene ribbons, 181, 194–95
graphene sheets, 10, 28–29, 84, 160, 192
graphene structures, 92, 182, 199
graphene superlattice (GSL), 164, 166–69
graphene synthesis, 15, 29, 32–37, 39, 42–43, 57
  trilayer, 41
graphene terraces, 132, 136
graphene transistors, 158
graphene trilayer, 44, 46, 49, 57, 61
graphene unit cell, 83, 93, 164, 168
graphite, 39–40, 61, 78, 93–94, 151, 183–88

natural, 155–56
pristine, 39
single-crystal, 2
single-layer, 182
thin, 183, 187
graphite enclosure, 184
graphite oxide, 2
graphitization, 12–13, 29–30, 32, 36–37, 79, 81, 84, 110, 116, 144, 188
metal-mediated, 14
nickel-mediated, 13
thermal, 79, 89
growth law, 124–27, 129–31, 135–36
cluster, 129
diffusion-type, 132
lateral, 129
linear, 133
parabolic, 135
power, 130
growth rate, 9, 79, 114, 117–19, 132–33, 135, 184, 188
GSL, *see* graphene superlattice
GSMBE, *see* gas-source molecular beam epitaxy

Hall bar, 193
Hall effect, quantum, 1, 189
Hall mobility value, low, 9
Hall plateau, 190
heating elements, 183–84
heteroepitaxial films, 11, 15
heteroepitaxy, 157
heterogeneous catalysts, 29
heterointerface, 163, 169
heterostack, 2D, 170
hexagonal symmetry, 93, 165
high-frequency transistor, 77, 189
highly oriented pyrolytic graphite (HOPG), 7–8, 28
high-resolution transmission electron microscopy (HRTEM), 156

high-temperature annealing, 40, 81–82, 89, 99, 110–11
H-intercalated graphene sample, 153
H-intercalated samples, 155
homogeneity, 147, 154, 161, 187, 190
honeycomb lattice, 29, 32, 37, 48–50, 166
honeycomb structure, 95
HOPG, *see* highly oriented pyrolytic graphite
HRTEM, *see* high-resolution transmission electron microscopy
hybridization, 75
hydrogen etching, 79, 94, 144–45
hydrogen intercalation, 144, 148
hydrogen passivation, 145
hydrogen treatment, 146, 149

IC, *see* integrated circuit
impurities, 37, 158
induction susceptor, 184
industrial microelectronic processing, 99
inhomogeneity, 147
integrated circuit (IC), 10, 31
integration, 130, 198–99
  direct, 15, 78
  large-scale, 181–82, 198–99
intensity ratio, 12, 112, 114
interaction
  electron–phonon, 158
  interatomic, 155–56
  tip–sample, 48, 51
  tip–surface, 49
intercalation, 9, 145–47, 149, 153, 158–61, 163, 165
  nitrogen, 160
  oxygen, 159
interface layer, 83–84, 110–11, 120–21, 142–44, 147, 158–59
interface structure, 33, 86, 89, 114

interfacial layer, 9, 33, 159
  2D ordered, 169
  carbon-rich, 9
interfacial order, 166
ion implantation, 6

jump-to-contact, 42
junctions, 159–60, 169, 190

kinetic models, 124, 132
kinetics, 116, 121, 125–26, 129, 131–32
kinetics of graphene growth, 109–10, 136
KJMA, *see* Kolmogorov–Johnson–Mehl–Avrami
KJMA equation, 132
Klein tunneling, 159–60
Knudsen cell, 163
Kolmogorov-Johnson-Mehl-Avrami (KJMA), 129

Landauer–Keldysh formalism, 61
Landau levels, 187
large-scale integration, 181–2, 198–9
laser beam, 4
laser-irradiated areas, 4–5
lateral superlattices, 164, 169
lattice constant, 89
lattice mismatch, 60, 142, 163, 165
lattices, 46, 48, 58, 80, 160, 164, 166
  graphene nanodomain, 58
law
  diffusional-growth, 136
  power, 116
layer-by-layer approximation, 123
layer-by-layer mode, 117–18
LDA, *see* local density approximation
LED, *see* light-emitting diode
LEED, *see* low-energy electron diffraction

LEED patterns, 37, 44, 46–47, 50, 86–88, 164
LEEM, *see* low-energy electron microscopy
LEEM micrograph, 44–45, 57, 148, 152, 154, 161–62
light-emitting diode (LED), 182
linear dispersions, 30, 52, 57, 63
line profile analysis, 147
local density approximation (LDA), 89–90
long-term annealing, 36
low-energy electron diffraction (LEED), 11, 35–38, 47, 50, 55, 65, 82, 84, 86, 98, 163, 166
low-energy electron microscopy (LEEM), 41, 44, 48, 55–56, 58, 84, 87–88, 96–97, 147–48, 161–63

magnetic field, 9, 190
magnetospectroscopy, 187
  infrared, 187
mask, 193
masking, 182, 194
mBZ, *see* mini Brillouin zone
MDC, *see* momentum distribution curve
MEG, *see* multilayered epitaxial graphene
MEMS, *see* microelectromechanical systems
microbeams, 14, 25
microdomains, 57
microelectromechanical systems (MEMS), 13, 157
micro-LEED, 44–46, 56–58
micromechanical exfoliation, 2
mini Brillouin zone (mBZ), 167–69
MLG, *see* monolayer graphene
mobilities, 8, 143, 145, 158, 187, 202
moiré patterns, 92–94, 185–86
moiré structure, 94

molecular beam epitaxy (MBE), 4, 7
momentum distribution curve (MDC), 155
monolayer (ML), 35, 44, 55–56, 88, 149, 161, 185, 195
monolayer graphene (MLG), 142–43, 146, 148–49, 153, 187–89
  as-grown, 143, 158
  freestanding, 48
  high-mobility, 190
  quasi-free, 142
  quasi-freestanding, 144
monolayers
  continuous, 110
  uniform, 79
monomethylsilane (MMS), 10
multilayered epitaxial graphene (MEG), 186–87, 189, 193
multilayer graphene, 6, 9, 11, 37, 81, 93, 98, 117, 193–94

nanodevices, 2
nanodomain boundary (NB), 42, 46, 52, 57–58, 60–63
nanodomains, 42–43, 46, 53–54, 57
  rotated, 63
  self-aligned, 60
nanoelectronics, 111, 182
nanoribbons, 78, 182
nanostructured graphene, 50, 57, 61–62
NB, *see* nanodomain boundary, 42, 46, 52, 57–58, 60–63
  self-aligned, 61–63
near-edge X-ray absorption fine structure (NEXAFS), 38–40, 55–56, 58
near-field microscopy (NFM), 189
NEXAFS, *see* near-edge X-ray absorption fine structure

NEXAFS spectra, 39–40, 57–58
NFM, *see* near-field microscopy
NMF, *see* nonnegative matrix factorization
non-Bernal stacking, 10, 33
nonnegative matrix factorization (NMF), 188
n-type doping, 143, 154
nucleation, 79, 117–18, 127, 129–31, 188

optical absorption dependence, 117
optical spectroscopy, 187
orbital structure, 50
orientation, 33, 47, 53–54, 80, 157, 164, 194–95
  armchair, 194
  domain, 47
  multicrystalline, 80
  preferential graphene lattice, 55
  preferential lattice, 37, 46, 57, 63
  uncontrolled, 192
  zigzag, 194
out-diffusion, 110, 116, 118, 120
overlayer, 40, 46, 50, 53, 78, 121–24, 131

passivation, 144, 172
peak FWHM cartography, 56
peak position cartography, 56
periodicity, 48–49, 60, 89–94, 142, 165–66, 168–69
periodic NBs, 60, 62
PES, *see* photoelectron spectroscopy
photoelectron diffraction, 114
photoelectrons, 50, 52, 114, 163
photoelectron spectroscopy (PES), 32, 35–36, 38–40
  angle-resolved, 30
  high-resolution X-ray, 111

photoemission, 40, 52, 62, 128, 155, 168
photolithography, 13, 198
photon energy, 35, 38–40, 45, 58, 115, 154, 164, 167
photonics, 15, 77, 96
photoresist, 13
photosensors, 27
planar channel material, 78
plane, basal, 194–95
plasma etching, 182, 191
plasmarons, 158, 168
PMMA, *see* poly(methylmethacrylate)
p/n junction, 159–60, 169
Poisson process, 129
Poisson statistics, 129
polarization, 146, 155, 157
polarized light, 40
polarized photons, 39, 58
polished substrates, 94, 96–97
polishing, 31, 79, 96, 191
polycrystalline graphene, suspended, 46
poly(methylmethacrylate) (PMMA), 4, 6
positive counterdoping, 189
power amplification, 190
p-type doping, 142, 146, 155, 159
pyroelectricity, 157

QFBLG, *see* quasi-freestanding bilayer graphene
QFMLG, *see* quasi-freestanding monolayer graphene
QFTLG, *see* quasi-freestanding trilayer graphene
QHE, *see* quantum Hall effect
quantization
  high-temperature, 189
  robust Hall resistance, 190
quantum Hall effect (QHE), 1, 189–90
quantum scattering, 53

Index | **215**

quasi-freestanding bilayer
graphene (QFBLG), 144,
149–51, 155
see quasi-freestanding monolayer
graphene (QFMLG), 144,
146–47, 155, 157–58
quasi-freestanding trilayer
graphene (QFTLG), 144,
154–57

radiation, 8, 182
synchrotron, 32, 175
radio frequency (RF), 184
Raman laser light, 188
Raman map, 193
Raman peaks, 4, 115, 188–89
Raman signals, 4, 40
Raman spectroscopy, 56, 81,
187–89
Raman spectrum, 4, 6, 34, 40–41,
55–56, 115, 188
rate constants, 126–27, 130–32
first-order, 125
reciprocal lattice vectors, 30,
163–65, 167
reciprocal vectors, 164
reconstructions, 35–38, 40, 54,
81–84, 89–90, 92–93, 110,
113, 163, 166, 169
carbon-rich, 142
intermediate, 92
reduction, 29, 190
chemical, 2
reflectivity, 148
reflectivity curves, 44
LEEM, 147
low-energy, 147
relative intensities, 35–36, 121
maximal, 39
reliability, 4, 13
resistance, 190, 197
length-independent, 197
tip-to-contact, 197

resolution, 42
high, 185
spatial, 147
submicron, 193
RF, *see* radio frequency
ribbon arrays, 181
ribbons, 194–99
ripples, 2, 48–49, 186
rippling, 2
atomic-scale, 48–49, 63
long-range, 151
RMS, *see* root mean square
root mean square (RMS), 42, 44,
185–86
rotational disorder, 84, 98
rotational variants, 52–53

sample areas, 55, 88, 154
large, 37
millimeter-scale, 50
sample heating, 35–36, 38
direct current, 35–36
scalable technique, 158
scanning electron microscopy
(SEM), 14, 186
scanning transmission electron
microscopy (STEM), 11,
94–95, 186
scanning tunneling microscopy
(STM), 11, 35–36, 38, 42, 44,
48, 50, 59, 81, 85, 89, 91, 94,
142, 147–48 165, 185–87
scanning tunneling spectroscopy
(STS), 187
scattering centers, 157
electronic, 190
SEM, *see* scanning electron
microscopy
semiconductor, 3, 13, 15, 29, 160
wide-bandgap, 29–30
semi-insulating substrate, 8
sensitivity, 13, 38
sensors, 77

SF, *see* stacking fault
sheets
   2D, 182
   continuous, 187
   extended monolayer, 188, 190
SiC chip, 184, 188, 199
sidewall ribbons, 197–98
sidewalls, 182, 185, 190–91,
   194–95, 198–99
silicon, 3, 8, 11–13, 15–16, 31, 36,
   40, 82, 125, 141, 159, 190,
   198–99
silicon carbide, 2–3, 12, 30–31
   heteroepitaxial, 15
   heteroepitaxial cubic, 3
silicon carbide substrates, low-cost
   single-crystal, 31
silicon carbide wafers, 15
silicon diffusion, 184
silicon sublimation, 183
silicon substrates, 13–15, 80
silicon wafers, 31–32
simulations, 52, 62, 158
   first-principles, 62
single-crystal ingots, 31
single-crystal metal surfaces, 29
single-crystal wafers, 32
single-layer graphene, 28, 88, 159,
   188
   freestanding, 30
Si-rich phase, 83, 86
Si-rich surface, 86, 112–13
solar cells, 3, 77
solution, wet chemical etch, 13–14
sp^2-bonded carbon atoms, 11
sp^2 hybridization, 36
sp^2-hybridized carbon atoms, 28
species
   chemisorbed, 146
   intercalated, 169
   reactive carbon, 124
spectroscopy, 11, 82, 145, 187
spectrum, 33, 35, 38, 40, 57, 111,
   121, 150, 154, 161, 168

core-level, 10, 33, 164
core-level photoelectron
   emission, 150
spin coating, 4
spin–orbit coupling, 160
spin transport, 187
stacking, 86, 114, 151–52, 156,
   170, 186, 199
   random, 186
   rhombohedral, 155
   rotational, 187
   rotational layer, 186
   turbostratic, 86
stacking fault (SF), 84, 97, 186
stacking order, 52–53
stacking sequences, 92, 151–52,
   155
STEM, *see* scanning transmission
   electron microscopy
step edges, 11, 42, 94–95, 110,
   117–18, 123, 147, 161, 190,
   194
step pinning, 191–92, 195
STM, *see* scanning tunneling
   microscopy
STM images, 42–44, 46–51, 54, 74,
   90–95, 113, 143, 146, 186
   atomically resolved, 46–47,
   58–59
   high-resolution, 47, 58, 93
   high-resolution atomic, 99
STM studies, 37, 42, 44, 46, 57, 84
stress-induced cracks, 82
structural disorder, 55
structural imperfections, 53
STS, *see* scanning tunneling
   spectroscopy
sublimation, 30, 36, 78, 96, 99,
   110, 116, 127, 130–33, 183–84
   confinement-controlled, 8, 79,
   182
   high-temperature, 88
   uniform, 96

sublimation rate, 9, 12, 79, 94, 96, 183

substrate intensity, 122–23, 131

substrates, 2–15, 28–29, 31–34, 40, 42–44, 46, 48–49, 53, 55, 78–85, 96–97, 120–23, 142–45, 165–66, 182
  as-grown, 97
  crystalline, 78
  large-area, 15
  off-axis, 94
  on-axis, 117
  ordered, 160
  single-crystal, 182
  transparent, 187
  unpolished, 96
  well-defined, 182

substrate spots, 37, 44, 47, 58

substrate surface, 7, 127
  defect-free, 96

supercapacitors, on-chip, 14

supercell, 142, 164–66

superlattice, 142, 160, 164–69

superlattice electron–phonon coupling, 160

superstructure, 164, 185
  moiré, 165

surface defects, 54, 127

surface density, 128, 130, 132, 134–35

surface diffusion, 132, 136

surface diffusion coefficient, 132–33

surface graphitization, 35, 40

surface reconstructions, 39, 41–42, 82–83, 85–86

surface termination, 33, 86, 114

synthesis, 28, 30–32, 34–35, 38, 54–55, 80
  direct, 29
  high-temperature, 31

TB, *see* tight binding

temperature cycle, typical, 184

temperature profile, 184–85

thermal annealing, 12, 110, 116, 136

thermal decomposition, 2, 7–10, 12, 30, 79, 110, 134, 183

thermal dissipation, 3

thermal equilibrium condition, 118

thermal inertia, 184

thermal mismatch, 80

thermal treatment, 191

three-layer graphene, 95–96

tight binding (TB), 52, 151–52, 154, 169

TLG, *see* trilayer graphene
  homogenous, 152
  rhombohedral, 152

top-gated transistors, 8

transformation, 78, 82, 89–92, 99, 126, 144

transition, 57, 90, 113, 126–27
  direct, 92
  progressive, 89
  superconducting, 160

transition metals, 2, 29, 158

transport gap, 54, 60–62, 64

trenches, 182, 194–95, 198–99
  parallel sidewall, 199
  shallow, 195

trilayer graphene (TLG), 45–46, 48–54, 56, 58, 60–61, 72, 142, 151–52, 155–56, 170, 174
  freestanding, 52
  nanostructured, 52, 61, 64
  uniform, 41, 56

tunneling currents, 50–51

tunneling gap, 50

UHV, *see* ultrahigh vacuum

ultrahigh vacuum (UHV), 7–10, 12, 28, 34–38, 40, 57–58, 78–82, 110–11, 114, 116–17, 120, 142, 144, 147–48, 197

unit cell, 10, 61, 83, 85, 91, 93, 142–43, 168

large, 166
lattice, 151
moiré, 165

vacancies, 55, 83, 125
vacuum, 29, 31, 86, 115, 183, 185, 191
vacuum chamber, 4, 37, 184
valence band, 31, 62, 98–99
vapor, Si, 116, 183, 188
vapor pressure, 184
vertical diffusion pathway, 118
Veselago lens, 160
vicinal sample, 57
viscous flow regime, 116
Voronoi cell, 134–35

wafer bonding, 198–99
wafers, 3–5, 10, 13, 15, 28–29, 31–37, 39, 45, 55–56, 60, 63, 181, 184, 191, 198–99
wafer scale, 30, 158, 182
wall
  cold, 145
  hot, 145

work function, 62
wrinkles, 92, 186

XPEEM, *see* X-ray photoelectron emission microscopy
XPS, *see* X-ray photoelectron spectroscopy
XPS spectra, 111–13
X-ray diffraction, 185–86
X-ray photoelectron emission microscopy (XPEEM), 162–63
X-ray photoelectron spectroscopy (XPS), 40, 81, 111–12, 114–15, 120–21, 126, 129, 136, 148, 150, 153

zero layer (ZL), 142, 144, 146, 157–58
zero-layer graphene (ZLG), 144, 146, 148, 153, 159, 161
zigzag structure, 62
ZL, *see* zero layer
ZLG, *see* zero-layer graphene
ZL image, 147
ZL structure, 148